AF787068

Direct Torque Control Strategies of Electrical Machines

Edited by Fatma Ben Salem

Published in London, United Kingdom

IntechOpen

Supporting open minds since 2005

Direct Torque Control Strategies of Electrical Machines
http://dx.doi.org/10.5772/intechopen.80103
Edited by Fatma Ben Salem

Contributors
Mohamed Haithem Lazreg, Abderrahim Bentaallah, Fatma Ben Salem, Yuting Gao, Yang Liu, Cherifi Djamila, Miloud Yahia, Dr. Venu Madhav Gopala, Obulesu Y. P., Adhavan Balashanmugham, Maheswaran Mockaisam, Sathiyanathan Murugesan, Tojiddin Khayrullaevich Juraev, Sobir Norov, Musulmanov Furqat Shodiyevich

© The Editor(s) and the Author(s) 2021
The rights of the editor(s) and the author(s) have been asserted in accordance with the Copyright, Designs and Patents Act 1988. All rights to the book as a whole are reserved by INTECHOPEN LIMITED. The book as a whole (compilation) cannot be reproduced, distributed or used for commercial or non-commercial purposes without INTECHOPEN LIMITED's written permission. Enquiries concerning the use of the book should be directed to INTECHOPEN LIMITED rights and permissions department (permissions@intechopen.com).
Violations are liable to prosecution under the governing Copyright Law.

Individual chapters of this publication are distributed under the terms of the Creative Commons Attribution 3.0 Unported License which permits commercial use, distribution and reproduction of the individual chapters, provided the original author(s) and source publication are appropriately acknowledged. If so indicated, certain images may not be included under the Creative Commons license. In such cases users will need to obtain permission from the license holder to reproduce the material. More details and guidelines concerning content reuse and adaptation can be found at http://www.intechopen.com/copyright-policy.html.

Notice
Statements and opinions expressed in the chapters are these of the individual contributors and not necessarily those of the editors or publisher. No responsibility is accepted for the accuracy of information contained in the published chapters. The publisher assumes no responsibility for any damage or injury to persons or property arising out of the use of any materials, instructions, methods or ideas contained in the book.

First published in London, United Kingdom, 2021 by IntechOpen
IntechOpen is the global imprint of INTECHOPEN LIMITED, registered in England and Wales, registration number: 11086078, 5 Princes Gate Court, London, SW7 2QJ, United Kingdom
Printed in Croatia

British Library Cataloguing-in-Publication Data
A catalogue record for this book is available from the British Library

Additional hard and PDF copies can be obtained from orders@intechopen.com

Direct Torque Control Strategies of Electrical Machines
Edited by Fatma Ben Salem
p. cm.
Print ISBN 978-1-83880-295-0
Online ISBN 978-1-83880-296-7
eBook (PDF) ISBN 978-1-83880-359-9

For EU product safety concerns:
IN TECH d.o.o., Prolaz Marije Krucifikse Kozulić 3, 51000 Rijeka, Croatia, info@intechopen.com or visit our website at intechopen.com.

We are IntechOpen, the world's leading publisher of Open Access books

Built by scientists, for scientists

5,100+
Open access books available

127,000+
International authors and editors

145M+
Downloads

156
Countries delivered to

Our authors are among the
Top 1%
most cited scientists

12.2%
Contributors from top 500 universities

CLARIVATE ANALYTICS
BOOK CITATION INDEX
INDEXED

WEB OF SCIENCE™

Selection of our books indexed in the Book Citation Index in Web of Science™ Core Collection (BKCI)

Interested in publishing with us?
Contact book.department@intechopen.com

Numbers displayed above are based on latest data collected.
For more information visit www.intechopen.com

Meet the editor

Fatma Ben Salem was born in Sfax, Tunisia, in 1978. She received her BS, MSc, Ph.D., and HCR degrees in 2002, 2003, 2010, and 2015, respectively, all in electrical engineering from the National Engineering School of Sfax, University of Sfax, Tunisia. She is an associate professor of electrical engineering at the High Institute of Industrial Management of Sfax, Tunisia. She is a member of the Control Energy Management Laboratory (CEMLab) of the University of Sfax. She is the author of several journals and international conference papers. She has participated in writing book chapters and in the organization of international conferences and workshops. She is an IEEE member. Her main research interests cover several aspects related to the control and the diagnostics of electrical machine drives and generators involved in automotive as well as in renewable energy systems.

Contents

Preface

Electrical machines are being widely used today in traction systems, robotics, and other position control systems. The modern applications demand advanced control techniques with high precision. The Direct Torque Control (DTC) method is one of the highest performance control strategies for AC machines to provide rapid torque and flux control. It is widely known to produce a quick and fast response in AC drives by selecting the proper voltage space vector according to the switching status of the inverter. The main advantage of DTC is its simple structure. Within this framework, this book entitled Direct Torque Control Strategies of Electrical Machines consists of a representation of theoretical results related to design and advanced control of AC machines and the diagnostics of electrical machine drives and generators involved in automotive systems. The priority has been focused on the DTC approach applications with and without commutation frequency control. It also covers DTC applications using artificial intelligence.

This book consists of seven chapters that have been written by leading researchers covering recent theoretical developments and applications of DTC strategies. It combines theoretical analysis, simulation, and experimental concepts. It is expected that readers require background knowledge to understand various concepts and results presented in this book.

Finally, the editor would like to express their sincere gratitude to authors from all over the world for submitting their high-quality work in a timely manner and revising it appropriately at short notice. A particular thanks and my deepest gratitude to the IntechOpen editorial staff for their continuous support, assistance, and significant improvement in the manuscript. Without their help, the book would not be published as scheduled.

Fatma Ben Salem
Associate Professor (ISGIS),
Sfax University, Tunisia

Section 1

Direct Torque Control of Induction Motors

Chapter 1

Improved Direct Torque Control Based on Neural Network of the Double-Star Induction Machine Using Deferent Multilevel Inverter

Mohamed Haithem Lazreg and Abderrahim Bentaallah

Abstract

In this chapter, we will compare the performance of a multilevel direct torque control (DTC) control for the double-star induction machine (DSIM) based on artificial neural network (ANN). The application of DTC control brings a very interesting solution to the problems of robustness and dynamics. However, this control has some disadvantages such as variable switching frequency, size, and complexity of the switching tables and the strong ripple torque. A solution to this problem is to increase the output voltage level of the inverter and associate the DTC control with modern control techniques such as artificial neural networks. Theoretical elements and simulation results are presented and discussed. As results, the flux and torque ripple of the five-level DTC-ANN control significantly reduces compared to the flux and torque ripple of the three-level DTC-ANN control. By viewing the simulation results using MATLAB/Simulink for both controls, the results obtained showed a very satisfactory behavior of this machine.

Keywords: double-star induction machine (DSIM), direct torque control (DTC), three-level inverter, five-level inverter, artificial neural network (ANN)

1. Introduction

The use of a conventional two-level inverter in the field of high power applications is not appropriate because it requires electronic components capable of withstanding high reverse voltage and high current. Another disadvantage of this inverter is the problem of magnetic interference caused by the abrupt change of the output voltage of the inverter from zero to high value [1].

With the appearance of the structures of the multilevel inverters proposed for the first time by [2], the research was able to face the handicaps presented by the classical structure. The goal of this research focus is to improve the quality of the output voltage, as well as to overcome the problems associated with two-level inverters. There are several topologies of multilevel inverters such as floating-diode, floating-capacitor, and cascaded inverters [3]. These structures make it possible to generate an output voltage of several levels.

Diode-clamped inverter (DCI) is the one that attracts the most attention because of the simplicity of its structure compared to the floating capacity inverter; in fact we do not need to use capacities for each phase, which eliminates the risks of parasitic resonances [4]. In this structure, diodes called floating diodes are associated with each phase, which serves to apply the different voltage levels of the DC source.

In high power, AC machines powered by static inverters find more and more applications. But the constraints on the power components limit the switching frequency and therefore the performance. To enable the use of higher switching frequency components, the power must be divided. To do this, one of the solutions is to use multiphase machines thanks to their advantages, such as the power segmentation and the minimization of the ripples of the torque (elimination of the harmonic torque of rank six). One of the most common examples of multiphase machines is the double-star induction machine (DSIM) [5].

To improve the decoupling between the flux and the torque, a so-called direct torque control (DTC) control technique has been applied.

The conventional direct torque control (DTCc) is proposed by Takahashi and Depenbrock in 1985 [2], and several studies allowed to apply this control technique on multiphase machines. As for each control, the DTC has advantages and disadvantages, and among these advantages, the stator resistance is theoretically the only parameter of the machine that intervenes in the control. This is essential for estimating the stator flux vector [6]. From this purely theoretical point of view, one can thus consider a great robustness compared to the other parameters of the machine; the block PWM is usually deleted [7].

Despite these advantages, this control also has significant disadvantages, the problem of instability such as the lack of control of the generator of acoustic noise at the machine. In addition, the use of hysteresis tapes is the cause of electromagnetic torque ripples and noise in the machine. To solve these drawbacks, in the framework of this work, we try to apply the multilevel direct torque control for DSIM and to develop a new control method such as artificial neural networks that replaces the switching tables [8].

This chapter is organized as follows: the DSIM model will be presented in the next section. The three-level and the five-level inverter modeling is described in the third and fourth section. The control method by DTC based on artificial neural networks (DTC-ANN) will be discussed in the fifth section. Moreover, in the sixth section, the simulation results are presented. Finally, a general conclusion summarizes this work.

2. DSIM model

In the conventional configuration, two identical three-phase windings share the same stator and are shifted by an electric angle of 30°. The rotor structure remains identical to that of a three-phase machine [9].

The model of machine DSIM is nonlinear. The DSIM model fed by voltage inverter is given by the following equations [10]:

$$\frac{dX}{dt} = AX + BU \tag{1}$$

$$T_{em} = p\frac{L_m}{L_r + L_m}\left[\varphi_{dr}(i_{qs1} + i_{qs2}) - \varphi_{qr}(i_{ds1} + i_{ds2})\right] \tag{2}$$

$$J\frac{d\Omega}{dt} = T_{em} - T_L - k_f\Omega \tag{3}$$

where:

$$X = [x_1, x_2, x_3, x_4, x_5, x_6]^T = \left[i_{ds1}, i_{ds2}, i_{qs1}, i_{qs2}, \phi_{dr}, \phi_{qr}\right]^T$$

$$U = \left[v_{ds1}, v_{ds2}, v_{qs1}, v_{qs2}\right]$$

Matrixes A and B are given by

$$A = \begin{bmatrix} a_1 & a_2 & a_3 & a_4 & a_5 & a_6 \\ -a_2 & a_1 & -a_4 & a_3 & -a_6 & a_5 \\ a_3 & a_4 & a_1 & a_2 & a_5 & a_6 \\ -a_4 & a_3 & -a_2 & a_1 & -a_6 & a_5 \\ a_9 & a_8 & a_7 & 0 & a_7 & 0 \\ -a_8 & a_9 & 0 & a_7 & 0 & a_7 \end{bmatrix} \qquad B = \begin{bmatrix} b_1 & 0 & b_2 & 0 \\ 0 & b_1 & 0 & b_2 \\ b_2 & 0 & b_1 & 0 \\ 0 & b_2 & 0 & b_1 \end{bmatrix}$$

where

$$a_1 = b_0\frac{L_m}{T_r} - b_1R_s,\ a_2 = \omega_s(b_1L_1 + b_2L_2),\ a_3 = b_0\frac{L_m}{T_r} - b_2R_s,\ a_4 = \omega_s(b_1L_2 + b_2L_1),$$

$$a_5 = \frac{-b_0}{T_r},\ a_6 = a_0b_3 + \omega_g b_0,\ a_7 = \frac{L_m}{T_r},\ a_8 = \omega_g,\ a_9 = -\frac{1}{T_r},\ a_{10} = \frac{3}{2}p\frac{L_m}{L_r},$$

$$\sigma = 1 - \frac{L_m^2}{L_sL_r},\ L_1 = \sigma L_s, L_2 = \sigma L_s - l_s, L_3 = L_s(1-\sigma),\ a_0 = \frac{L_m}{L_r},\ b_0 = \frac{L_m}{L_r(L_1 + L_2)},$$

$$b_1 = \frac{L_1}{L_1^2 - L_2^2},\ b_2 = \frac{L_2}{L_1^2 - L_2^2},\ b_3 = \omega_s(b_1 + b_2)$$

3. Modeling of three-level inverter

Figure 1 shows the structure of the three-level floating-diode inverter introduced by A. Nabae and H. Akagi in 1981 [11] (**Table 1**).

The three symmetrical arms consist of four fully controllable switches. These switches must not be opened or closed simultaneously, in order to avoid short circuiting of the DC source at the input of the inverter. Each switch is composed of an antiparallel transistor with a diode. The floating diodes ensure the application of the different voltage levels at the output of each arm. The DC input voltage is divided into two equal parts by using two capacitors. Each capacitor must be sized for a voltage equal to $v_{dc}/2$ [12].

The switching function of each switch T_{xki} (k = 1, 2, i = 1 ... 4, x = a, b, and c) is defined as follows:

$$F_{xki} = \begin{cases} 1\ if\ T_{xki}\ is\ ON \\ 0\ if\ T_{xki}\ is\ OFF \end{cases} \tag{4}$$

The controls of the switches of the lower half-arms are complementary to those of the upper half-arms:

$$F_{xki} = 1 - F_{xk(i-2)} \tag{5}$$

For each arm, we define three connection functions:

$$\begin{cases} F_{c1xk} = F_{c1xk}F_{c2xk} \\ F_{c2xk} = F_{c2xk}F_{c3xk} \\ F_{c3xk} = F_{c3xk}F_{c4xk} \end{cases} \tag{6}$$

The output voltages with respect to the neutral point of the DC source are expressed by

$$\begin{pmatrix} v_{aOk} \\ v_{bOk} \\ v_{cOk} \end{pmatrix} = \begin{pmatrix} F_{c1ak} & F_{c2ak} & F_{c3ak} \\ F_{c1bk} & F_{c2bk} & F_{c3bk} \\ F_{c1ck} & F_{c2ck} & F_{c3ck} \end{pmatrix} \begin{pmatrix} v_{c2} \\ 0 \\ -v_{c1} \end{pmatrix} \tag{7}$$

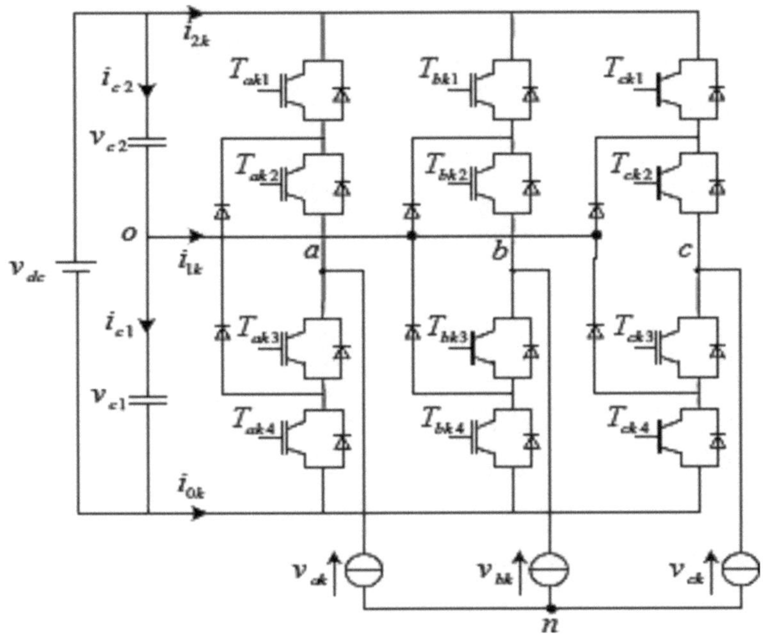

Figure 1.
Three-phase inverter with floating diodes (k = 1 is the first inverter, and k = 2 is the second inverter).

Switching states	State of the switches of an arm				Output voltage
	T_{xk1}	T_{xk2}	T_{xk3}	T_{xk4}	
2	1	1	0	0	v_{c2}
1	0	1	1	0	0
0	0	0	1	1	$-v_{c1}$

Table 1.
States of an arm of the inverter with three levels.

4. Modeling of five-level inverter

Currently the diode-clamped inverter is the one that attracts the most attention, given the simplicity of its structure compared to floating capacity inverters and cascading. In fact, compared to the inverter with floating capacities, it is not necessary to use capacities for each phase, which eliminates the risks of parasitic resonances.

The main advantage lies in a considerable reduction in switching losses and its ability to control harmonic content [13].

Figure 2 shows the structure of the inverter with five levels, each of the three arms of the inverter consists of eight controlled switches and six floating diodes. The controlled switches are unidirectional in voltage and bidirectional current; it is conventional associations of a transistor and an antiparallel diode.

These switches must not be opened or closed simultaneously, in order to avoid a short circuit of the DC source in the input. The floating diodes (six per arm) ensure the application of the different voltage levels at the output of each arm. The DC input voltage is divided into four equal parts using four capacitors [14].

The DC input bus is composed of four capacitors (C1, C2, C3, and C4), making it possible to create a set of three capacitive middle points. The total voltage of the DC bus is v_{dc}; under normal operating conditions, this is uniformly distributed over the four capacitors, which then have a voltage $v_{dc}/4$ at their terminals [15] (**Table 2**).

For each switch T_{xki} (k = 1, 2, i = 1 ... 8, x = a, b, and c), a switching function is defined as follows:

$$F_{xki} = \begin{cases} 1 \text{ if } T_{xki} \text{ is ON} \\ 0 \text{ if } T_{xki} \text{ is OFF} \end{cases} \tag{8}$$

The switch control of the lower half-arms is complementary to those of the upper half-arms:

$$F_{xki} = 1 - F_{xk(i-4)} \tag{9}$$

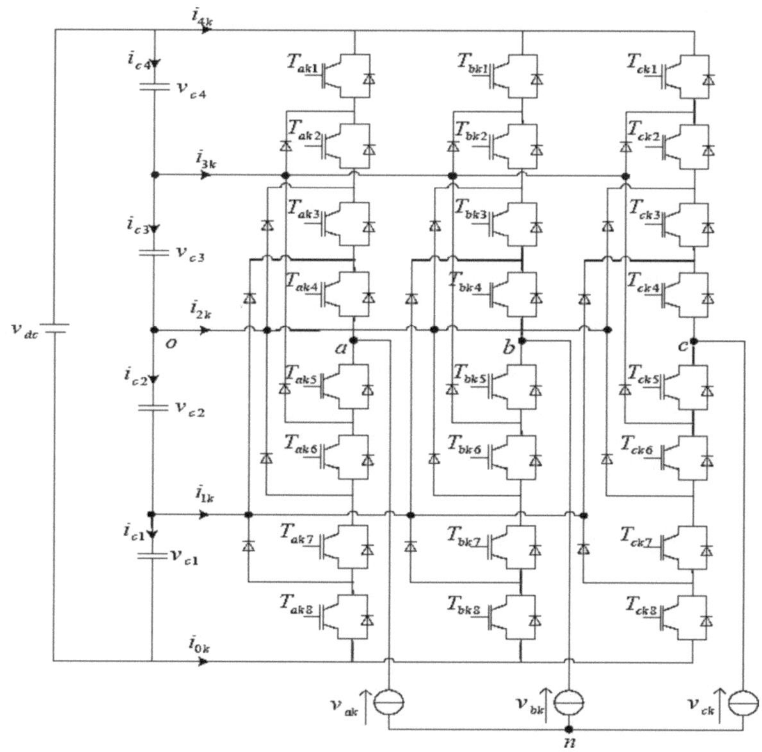

Figure 2.
Diagram of the five-level inverter with NPC structure.

Switching states	State of the switches of an arm								Output voltage
	T_{xk1}	T_{xk2}	T_{xk3}	T_{xk4}	T_{xk5}	T_{xk6}	T_{xk7}	T_{xk8}	
4	1	1	1	1	0	0	0	0	$V_{c3} + V_{c4}$
3	0	1	1	1	1	0	0	0	V_{c3}
2	0	0	1	1	1	1	0	0	0
1	0	0	0	1	1	1	1	0	$-V_{c2}$
0	0	0	0	0	1	1	1	1	$-(V_{c1}+ V_{c2})$

Table 2.
States of an arm of the inverter with five levels.

We define five connection functions, each associated with one of the five states of the arm:

$$\begin{cases} F_{c1xk} = F_{c1xk}F_{c2xk}F_{c3xk}F_{c4xk} \\ F_{c2xk} = F_{c2xk}F_{c3xk}F_{c4xk}F_{c5xk} \\ F_{c3xk} = F_{c3xk}F_{c4xk}F_{c5xk}F_{c6xk} \\ F_{c4xk} = F_{c4xk}F_{c5xk}F_{c6xk}F_{c7xk} \\ F_{c5xk} = F_{c5xk}F_{c6xk}F_{c7xk}F_{c8xk} \end{cases} \tag{10}$$

The potentials of nodes a, b, and c of the three-phase inverter at five levels with respect to the point o are given by the following system:

$$\begin{pmatrix} v_{a0k} \\ v_{b0k} \\ v_{c0k} \end{pmatrix} = \begin{pmatrix} F_{c1ak} & F_{c2ak} & F_{c3ak} & F_{c4ak} & F_{c5ak} \\ F_{c1bk} & F_{c2bk} & F_{c3bk} & F_{c4bk} & F_{c5bk} \\ F_{c1ck} & F_{c2ck} & F_{c3ck} & F_{c4ck} & F_{c5ck} \end{pmatrix} \begin{pmatrix} v_{c3} + v_{c4} \\ v_{c3} \\ 0 \\ -v_{c2} \\ -(v_{c1} + v_{c2}) \end{pmatrix} \tag{11}$$

5. Direct torque control based on neural networks

The direct torque control of a DSIM is based on the direct determination of the control sequence applied to the switches of a voltage inverter. This choice is based generally on the use of hysteresis comparators whose function is to control the state of the system, namely, the amplitude of the stator flux and the electromagnetic torque [16].

In the structure of the DTC, the voltage model is commonly used. Thus, the amplitude of the stator flux is estimated from its components following the axes (α, β):

$$\begin{cases} \hat{\phi}_{\alpha s} = \int_0^t (V_{\alpha s} - R_s I_{\alpha s}) dt \\ \hat{\phi}_{\beta s} = \int_0^t (V_{\beta s} - R_s I_{\beta s}) dt \end{cases} \tag{12}$$

The stator flux module is given by

$$\hat{\varphi}_s = \sqrt{\hat{\varphi}_\alpha^2 + \hat{\varphi}_\beta^2} \tag{13}$$

The angle θ_s is given by

$$\hat{\theta}_s = \tan 2^{-1}\left(\frac{\hat{\varphi}_\beta(t)}{\hat{\varphi}_\alpha(t)}\right) \tag{14}$$

This method of estimating the stator flux has the advantage of simplicity and accuracy, particularly at medium and high speeds where the ohmic voltage drop becomes negligible [17].

The electromagnetic torque can be estimated from the estimated magnitudes of the stator flux and the measured magnitudes of the line currents, by the following equation:

$$\hat{T}_{em} = \frac{3}{2}p.\left(\hat{\phi}_{\alpha s} i_{\beta s} - \hat{\phi}_{\beta s} i_{\alpha s}\right) \tag{15}$$

5.1 Neural network strategy

The human brain is able to adapt, learn, and decide, and it is on this fact that researchers have been interested in understanding its operating principle and being able to apply it to the field of computer science.

Among the disadvantages of DTC control, a slow response for small changes in stator flux and electromagnetic torque, size, and complexity of switching tables when the number of levels of inverters is high. In order to improve the performance of the DTC control, many contributions have been made in the DTC control based on artificial neural networks [18].

In this application, our goal is to replace switching tables with artificial neural networks.

The multilayer architecture was chosen to be applied to multilevel DTC control. This network, which can be multiplexed for each controller output, has acceptable performance in many industrial applications [19]. The neural network contains three layers: input layer, hidden layers, and output layer. Each layer consists of several neurons. The number of neurons in the output and the layers depends on the number of input and output variables chosen. The number of hidden layers and the number of neurons in each one depend on the dynamics of the system and the desired degree of accuracy.

Figure 3 shows the structure of the neural network applied to the multilevel DTC control of the DSIM. It is a network with three neurons in the input layer, whose inputs are flow error (Ef), torque error (Ec), and flow position angle (Z) [20]. For the three-level inverter, there are 12 neurons in the hidden layer and 06 neurons in the output, and for the five-level inverter, there are 24 neurons in the hidden layer and 12 neurons in the output. **Figure 4** shows the chosen architecture.

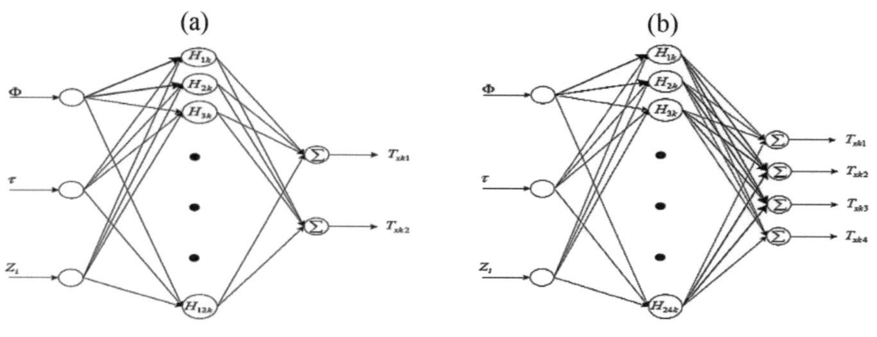

Figure 3.
Neural network structure applied to the multilevel DTC control. (a) for three-level DTC, (b) for five-level DTC.

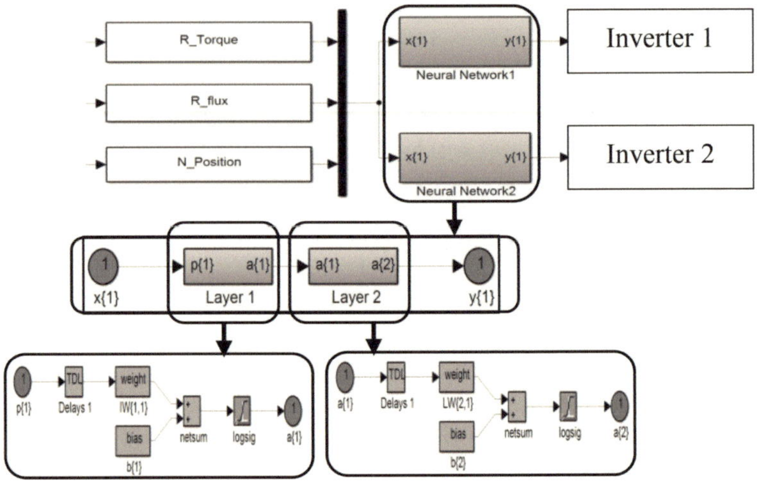

Figure 4.
Selection table based on neuron network.

6. Simulation results

In order to test the static and dynamic performance of the control, the DSIM is accelerated from standstill to reference speed 100 rad/s. The machine is applied

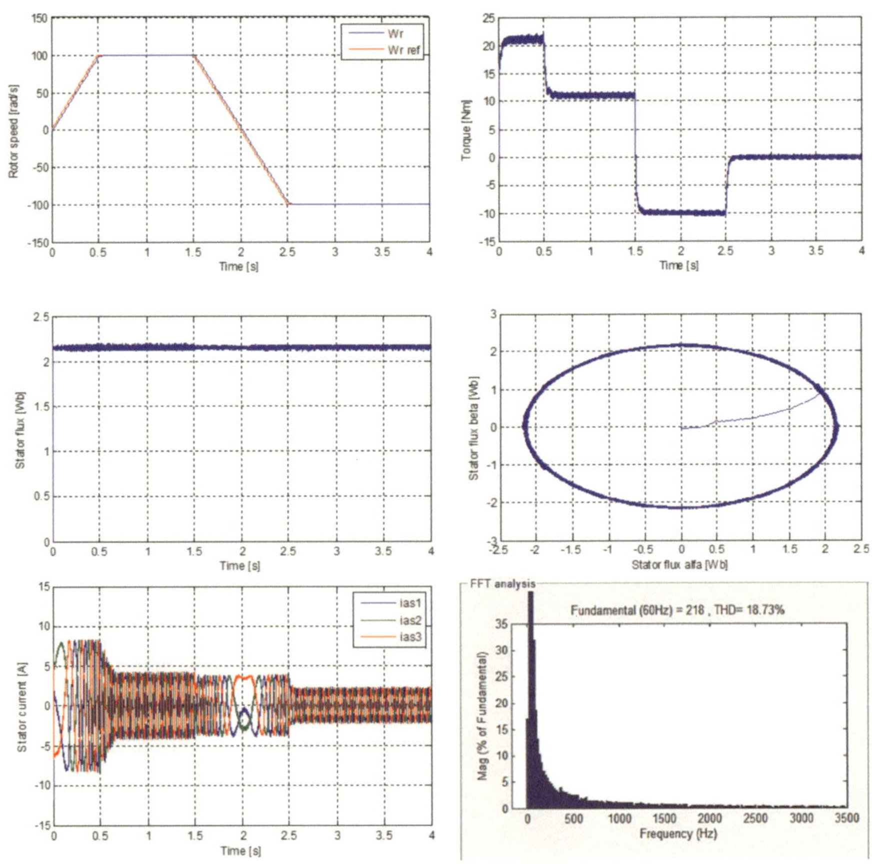

Figure 5.
Simulation results of real and estimated speed, torque, flux, and current of three-level DTC-ANN.

with a load torque of 11 Nm. Finally, the direction of rotation of the machine is reversed from 100 rad/s to −100 rad/s at time t = 2 s. **Figures 5** and **6** show the simulation results of the three- and five-level DTC control for DSIM.

Simulation results of speed, stator flux, torque, stator current, and stator voltage show the good performance of the three- and five-level DTC-ANN control of DSIM (speed, stability, and precision).

We note that the speed follows its reference value. The electromagnetic torque stabilizes at the value of the nominal torque after a transient regime with rapid response and without exceeding before stabilizing at the value of the applied load torque.

Figure 6 shows that the five-level DTC-ANN control reduces the ripple of the electromagnetic torque, the stator flux, and the THD value compared to that of the three-level DTC-ANN. On the other hand, we note that the speed reaches its reference without exceeding for the two control types. Moreover, the couple follows the load torque. The dynamics of the stator flux are not affected by the application of these load instructions.

The use of multilevel inverter at five levels causes a decrease in the current ripple at the steady state that is to say low peaks than that of the three-level control. However, the results of the simulations shows a good dynamic characteristic of the stator flux in the transient regime for five-level DTC-ANN compared to the three-level DTC-ANN with static errors that are virtually null in both cases of control DTC proposed.

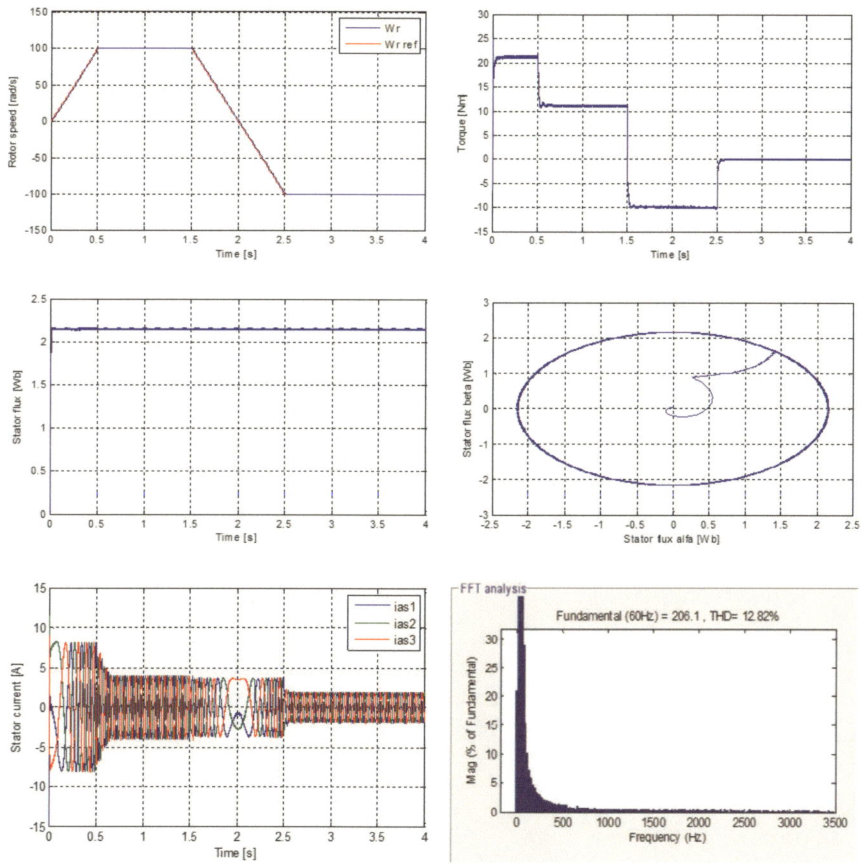

Figure 6.
Simulation results of real and estimated speed, torque, flux, and current of five-level DTC-ANN.

Figures 7 and **8** show the simulation results of the three-level and five-level DTC-ANN control for low-speed operation. DSIM is accelerated from standstill to a low reference speed of 10 rad/s, at time t = 0.5 s; the DSIM is accelerated again to a reference speed of 100 rad/s. The machine is loaded with a nominal load of 11 Nm. Finally, a reversal of the direction of rotation of the machine from 100 rad/s to −10 rad/s is performed at time t = 2 s.

The simulation results show that low-speed operation does not affect the performance of the proposed drive. Indeed, the good reference speed tracking is ensured, with advantages brought by the use of five-level DTC-ANN control, the minimization of torque ripple, and stator flux, which is confirmed by the simulation results.

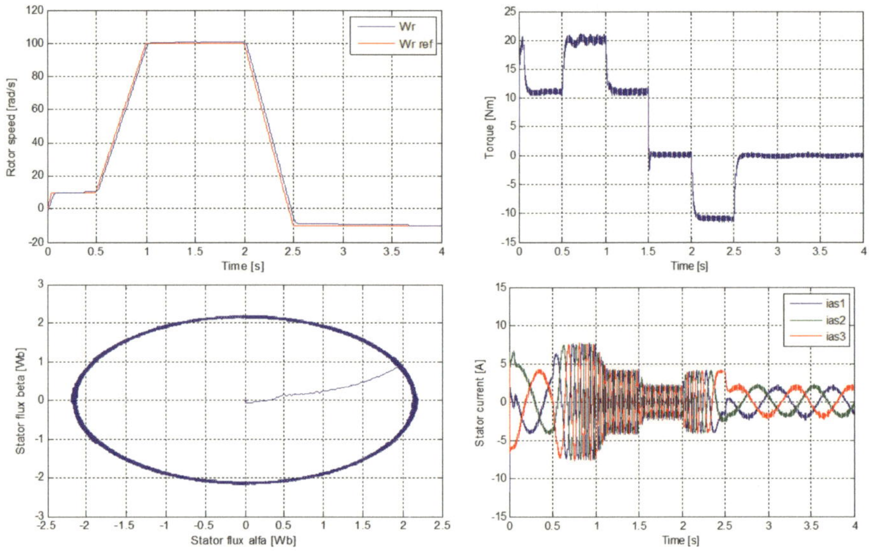

Figure 7.
Simulation results of three-level DTC-ANN for low-speed operation.

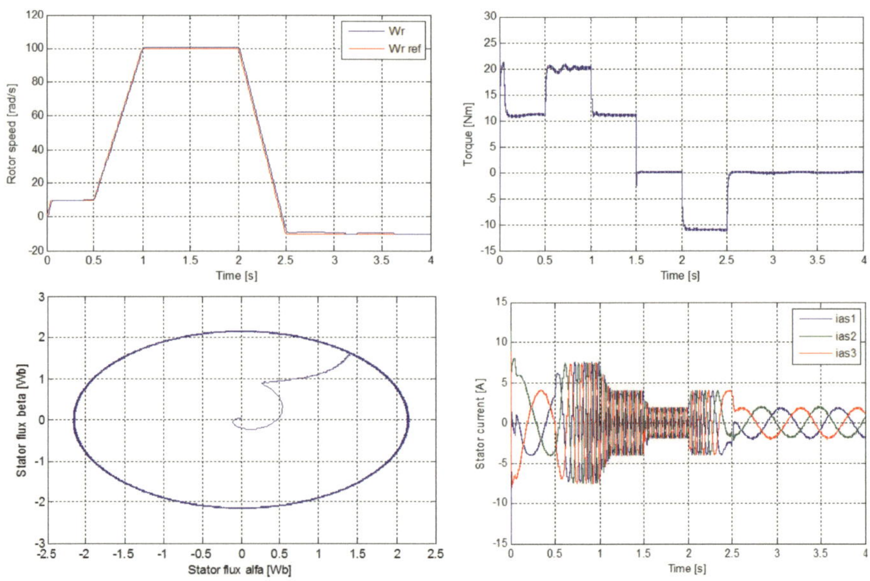

Figure 8.
Simulation results of five-level DTC-ANN for low-speed operation.

	THD (%)	Ripples of torque	Ripples of flux
Three-level DTC	18.73	Good	Good
Five-level DTC	12.82	Very good	Very good

Table 3.
The comparison between three-level and five-level DTC-ANN.

In order to know the best type control of DSIM, a comparative study is essential between the two types (three-level DTC-ANN and five-level DTC-ANN). The following table shows the comparison between the two types (**Table 3**).

7. Conclusion

In this chapter, we presented two types of DTC control (three-level DTC-ANN and five-level DTC-ANN) of a DSIM fed by two NPC voltage inverters, and the technique of neural networks was applied to the DTC control. The main advantage of this control is to allow control of the flux and torque of the machine without the need to use a mechanical sensor. The direct torque control strategy is an effective and simple way to control an induction machine. In order to improve the performance of the DSIM (torque ripple reductions, flux, response time, and the THD value of the stator current), simulation tests of the control by variation and inversely of the speed have been presented; the results obtained show that the five-level DTC-ANN control with speed control is very efficient. This shows the effectiveness of the proposed strategy.

Appendix

DSIM parameters

P_n = 4.5 Kw
I_n = 6 A
R_r = 2.12 Ω
L_r = 0.006 H
R_{s1} = R_{s2} = 1.86 Ω
L_{s1} = L_{s2} = 0.011 H
L_m = 0.3672 H
J = 0.065 kg.m^2
k_f = 0.001 Nm/rad.

Author details

Mohamed Haithem Lazreg* and Abderrahim Bentaallah
Faculty of Electrical Engineering, Laboratory of Intelligent Control and Electrical power Systems (ICEPS), Djillali Liabes University of Sidi Bel Abbes, Algeria

*Address all correspondence to: haitem.31@hotmail.fr

IntechOpen

© 2020 The Author(s). Licensee IntechOpen. This chapter is distributed under the terms of the Creative Commons Attribution License (http://creativecommons.org/licenses/by/3.0), which permits unrestricted use, distribution, and reproduction in any medium, provided the original work is properly cited.

References

[1] Kirankumar B, Reddy YV, Vijayakumar M. Multilevel inverter with space vector modulation: Intelligence direct torque control of induction motor. IET Power Electronics. 2017;**10**:1129-1137

[2] Nabae A, Takahashi I, Akagi H. A new neutral-point-clamped PWM inverter. IEEE Transactions on Industrial Applications. 1981;**IA-17**: 518-523

[3] Babaei E, Dehqan A, Sabahi M. A new topology for multilevel inverter considering its optimal structures. Electric Power Systems Research. 2013; **103**:145-156

[4] Gupta K, Bhatnagar P. Chapter 2 – Basics of multilevel inverters. In: Multilevel Inverters Conventional and Emerging Topologies and their Control; 2018. pp. 21-42

[5] Meroufel A, Massoum S, Bentaallah A, et al. Double star induction motor direct torque control with fuzzy sliding mode speed controller. Revue Roumaine des Sciences Techniques - Serie Électrotechnique et Énergétique. 2017; **62**:31-35

[6] Taheri A. Harmonic reduction of direct torque control of six-phase induction motor. ISA Transactions. 2016;**63**:299-314

[7] Khedher A, Mimouni MF. Sensorless-adaptive DTC of double star induction motor. Energy Conversion and Management. 2010;**51**:2878-2892

[8] Gadoue SM, Giaouris D, Finch JW. Artificial intelligence-based speed control of DTC induction motor drives — A comparative study. Electric Power Systems Research. 2009;**79**:210-219

[9] Lazreg MH, Bentaallah A. Speed sensorless vector control of double star induction machine using reduced order observer and MRAS estimator. In: IEEE International Conference on Electrical Engineering (ICEE-B); Boumerdes, Algeria; 2017

[10] Lazreg MH, Bentaallah A. Sensorless fuzzy sliding-mode control of the double-star induction motor using a sliding-mode observer. Elektrotehniški Vestnik. 2018;**85**:169-176

[11] Payami S, Behera RK, Iqbal A. DTC of three-level NPC inverter fed five-phase induction motor drive with novel neutral point voltage balancing scheme. IEEE Transactions on Power Electronics. 2018;**33**:1487-1500

[12] Harina MM, Vanithaa V, Jayakumar M. Comparison of PWM techniques for a three level modular multilevel inverter. Energy Procedia. 2017;**117**:666-673

[13] Figarado S, Sivakumar K, Ramchand R, Das A, Patel C, Gopakumar K. Five-level inverter scheme for an open-end winding induction machine with less number of switches. IET Power Electronics. 2010;**3**: 637-647

[14] Rosmadi A, Nasrudin A, Siti R, Sheikh R, Zaharin A. A five-level diode-clamped inverter with three-level boost converter. IEEE Transactions on Industrial Electronics. 2014;**61**: 5155-5163

[15] Naganathan P, Srinivas S, Ittamveettil H. Five-level torque controller-based DTC method for a cascaded three-level inverter fed induction motor drive. IET Power Electronics. 2017;**10**:1223-1230

[16] Talaeizadeh V, Kianinezhad R, Seyfossadat SG, Shayanfar HA. Direct torque control of six-phase induction motors using three-phase matrix

converter. Energy Conversion and Management. 2010;**51**:2482-2491

[17] Usta MA, Okumus HI, Kahveci HA. Simplified three-level SVM-DTC induction motor drive with speed and stator resistance estimation based on extended Kalman filter. Electrical Engineering. 2017;**99**:707-720

[18] Barika S, Jaladi KK. Five-phase induction motor DTC-SVM scheme with PI controller and ANN controller. Procedia Technology. 2016;**25**:816-823

[19] Zolfaghari M, Taher SA, Munuz DV. Neural network-based sensorless direct power control of permanent magnet synchronous motor. Ain Shams Engineering Journal. 2016;7:729-740

[20] Douiri MR, Cherkaoui M. Comparative study of various artificial intelligence approaches applied to direct torque control of induction motor drives. Frontiers in Energy. 2013;7: 456-467

Chapter 2

Direct Torque Control Strategies of Induction Machine: Comparative Studies

Cherifi Djamila and Miloud Yahia

Abstract

The direct torque control (DTC) was proposed as an alternative to the vector control in the middle of 1980s for AC machine control. This strategy bases on the direct determination of inverter switching states and offers a simpler scheme and less sensitivity to machine parameters. However, the variable switching frequency of DTC causes high flux and torque ripples which lead to an acoustical noise and degrade the performance of the control technique, especially at low-speed regions. In the objective of improving the performance of DTC for the induction motor, this work addresses the most important points concerning this issue. The reduction of high ripples, which are the major drawbacks, by applying a constant switching frequency using the space vector modulation (SVM) has been done firstly. Then, fuzzy DTC-SVM strategy with adaptive fuzzy-PI speed controller has been proposed. The results of all the discussed aspects of this chapter have been obtained by numerical simulation using MATLAB/Simulink software.

Keywords: induction motor, conventional direct torque control (DTC), space vector modulation, DTC-SVM, fuzzy-PI speed controller

1. Introduction

Advanced control of electrical machines requires an independent control of magnetic flux and torque. For that reason it was not surprising that the DC machine played an important role in the early days of high-performance electrical drive systems, since the magnetic flux and torque are easily controlled by the stator and rotor current, respectively. The introduction of field oriented control meant a huge turn in the field of electrical drives, since with this type of control the robust induction machine can be controlled with a high performance. Later in the 1980s, a new control method for induction machines was introduced: The direct torque control (DTC) method. It was proposed by Takahashi and Depenbrock [1, 2]. It bases on the direct selecting of the switching states to control the voltage source inverter (VSI) through a switching look-up table. Due to the limits of the conventional DTC strategy, especially the high torque and flux ripples problem, various control structures are presented to improve the performances of control, [3, 4]. The constant switching frequency DTC using the space vector modulation (DTC-SVM) is a well discussed solution; in order to improve the DTC-SVM performances, hysteresis comparators of electromagnetic torque and stator flux have been

 IntechOpen

replaced by PI controllers, [5, 6]. The main drawbacks of DTC-SVM using PI controllers are the sensitivity of the performances to the system-parameter variations and the inadequate rejection of external disturbances and load changes [7–11]. To cope with this disadvantage, it is suggested to replace the conventional regulators used for the speed control, flux, and electromagnetic torque by intelligent controllers by fuzzy logic to make the controls more robust against the disturbances of the parameters of the machine. The aim of this chapter is to design and compare three strategies for the direct torque control (DTC) of induction motor (IM). The first method is a conventional direct torque control (C-DTC) where the torque and the flux are regulated by the hysteresis controllers. The second one is direct torque control by space vector modulation strategy (SVM-DTC) where the torque and flux are regulated by PI controllers. The third one is fuzzy SVM-DTC with adaptive fuzzy-PI speed controller where the torque and flux are regulated by fuzzy logic controllers. The main feature of the proposed (SVM-DTC) strategy is the reduction of torque and flux ripples.

2. Model of induction motor dedicated for direct torque control

The mathematical model of induction motor can be described in the stator fixed reference frame (α,β) (stationary frame) by assuming the rotor and the stator flux as state variables:

$$\frac{dX}{dt} = AX + BU \tag{1}$$

with

$$X = \begin{bmatrix} \phi_{s\alpha} \\ \phi_{s\beta} \\ \phi_{r\alpha} \\ \phi_{r\beta} \end{bmatrix},\ A = \begin{bmatrix} -\frac{R_s}{\sigma L_s} & 0 & \frac{MR_s}{oL_rL_s} & 0 \\ 0 & -\frac{R_s}{\sigma L_s} & 0 & \frac{MR_s}{oL_rL_s} \\ \frac{MR_r}{oL_rL_s} & 0 & -\frac{R_r}{\sigma L_r} & -\omega \\ 0 & \frac{MR_r}{oL_rL_s} & \omega & -\frac{R_r}{\sigma L_r} \end{bmatrix},\ B = \begin{bmatrix} 1 & 0 \\ 0 & 1 \\ 0 & 0 \\ 0 & 0 \end{bmatrix} \text{ and}$$

$$U = \begin{bmatrix} v_{s\alpha} \\ v_{s\beta} \end{bmatrix} \text{ where}$$

$\phi_{s\alpha}$, $\phi_{s\beta}$, $\phi_{r\alpha}$, $\phi_{r\beta}$ are stator and rotor flux components.
R_s, R_r are stator and rotor resistance.
L_s, L_r are stator and rotor inductance.
M is the mutual stator-rotor inductance.
$\sigma = 1 - \frac{M^2}{L_rL_s}$ is the Blondel's coefficient.
ω is the machine speed ($\omega = p\Omega = \omega_s$-$\omega_r$ and p is the pole pair number).

The rotor motion can be described by:

$$J\frac{d\Omega}{dt} = T_{em} - T_L - f\,\Omega \tag{2}$$

where J is the motor inertia, T_{em} is the electromagnetic torque, T_L is the load torque, and f is the friction coefficient.

3. Two-level voltage source inverter (VSI) model

Two-level three-phase voltage source inverter (VSI) is considered as a mature technology and becoming an industrial standard for the demand for energy saving. The output phase voltages are produced by the rectifier (V_{dc}) is delivered to the inverter input, which, thanks to controlled transistor switches, converts this voltage to three-phase AC voltage signal with wide range variable voltage amplitude and frequency.

The type of the used switches depends on the power of the inverter and switching frequency. In the most applications, IGBT transistors with antiparallel diodes are so helpful.

The model of two-level voltage inverter is shown in **Figure 1**.

Figure 1 shows the two-level three-phase voltage source inverter (VSI) with six transistor switches, S1–S6, and a dc constant voltage source V_{dc} connecting a three-phase load.

The voltage vector is generated by the following equation:

$$\vec{V}_s = \sqrt{\frac{2}{3}} V_{dc} \left[S_a + S_b e^{j\frac{2\pi}{3}} + S_c e^{j\frac{4\pi}{3}} \right] \tag{3}$$

where S_a, S_b, and S_c are three-phase inverter switching functions, which can take a logical value of either 0 or 1.

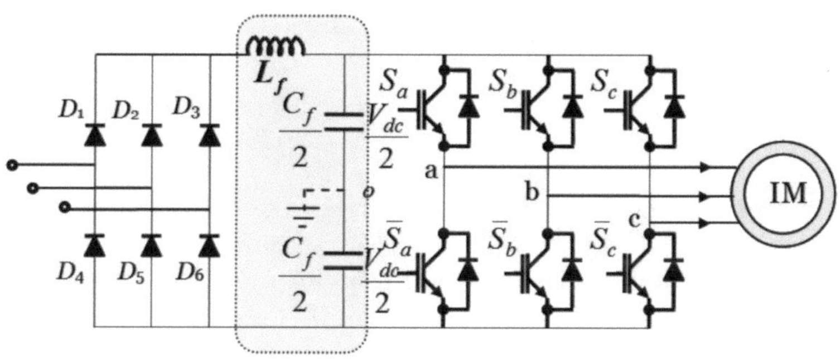

Figure 1.
Three-phase VSI fed star-connected induction machine.

4. Principles of direct torque control

Direct torque control principle was introduced in the late 1980s by [1, 2]. It achieves a decoupled control of the stator flux and the electromagnetic torque in the stationary frame (α, β), and it allows induction machines to have an accurate and fast electromagnetic torque response. It uses a switching table for the selection of an appropriate voltage vector. The selection of the switching states is related directly to the variation of the stator flux and the torque of the machine. Hence, the selection is made by restricting the flux and torque magnitudes within two hysteresis bands. Those controllers ensure a separated regulation of both of these quantities [12–14]. The inputs of hysteresis controllers are the flux and the torque errors as well as their outputs determine the appropriate voltage vector for each commutation period.

5. Estimation of stator flux and electromagnetic torque

5.1 Control of stator flux

Basing on the induction motor model in stationary frame, the stator flux equation can be expressed as follows [15–19]:

$$\begin{cases} \phi_{s\alpha} = \int (v_{s\alpha} - R_s i_{s\alpha}) dt \\ \phi_{s\beta} = \int \left(v_{s\beta} - R_s i_{s\beta}\right) dt \end{cases} \tag{4}$$

Considering that the control of the switches of the inverter is done by control period (or sampling) T_e and that at each of these periods the states S_a, S_b, and S_c are kept constant, the method of numerical integration of the rectangles makes it possible to obtain an expression of the sample *k + 1* of the stator flux in the following form:

$$\begin{cases} \phi_{s\alpha}(k+1) = \phi_{s\alpha}(k) + (v_{s\alpha}(k) - R_s i_{s\alpha}(k))\, T_e \\ \phi_{s\beta}(k+1) = \phi_{s\beta}(k) + \left(v_{s\beta}(k) - R_s i_{s\beta}(k)\right)\, T_e \end{cases} \tag{5}$$

A vector inscription of this expression can be given by:

$$\vec{\phi}_s(k+1) = \vec{\phi}_s(k) + \left(\vec{V}_s(k) - R_s \vec{I}_s(k)\right) T_e \tag{6}$$

We can neglect the stator resistance voltage drop compared to *Vs* for high speed regions. Then Eq. (6) can be written as:

$$\vec{\phi}_s(k+1) = \vec{\phi}_s(k) + \vec{V}_s(k) T_e \tag{7}$$

Eq. (7) means that the stator flux can be changed by the application of stator voltage during a time *k*. The stator flux vector's extremity moves in direction given by the voltage vector and making a circular trajectory.

A two-level hysteresis comparator is used for flux regulation. It allows to drop easily the flux vector extremity within the limits of the two concentric circles with close radius. The choice of the hysteresis bandwidth depends on the switching frequency of the inverter **Figures 2** and **3**.

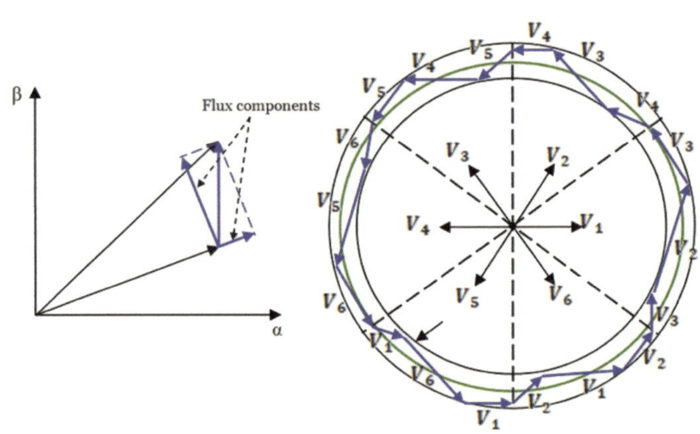

Figure 2.
Evolution of stator flux vector in the complex plan.

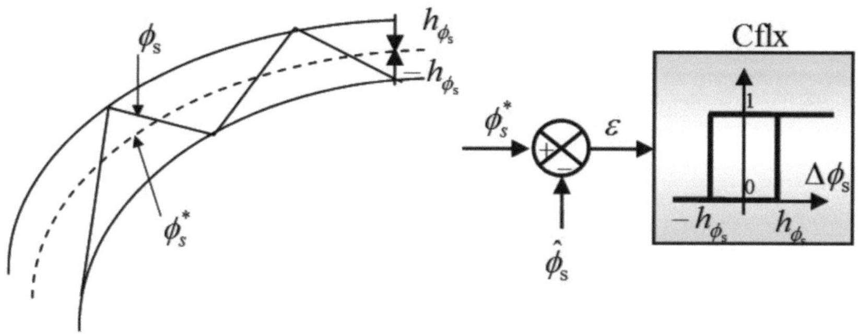

Figure 3.
Two-level hysteresis comparator for stator flux control.

The logical outputs of the flux controller are defined as:

$$\begin{aligned} &\text{Cflx} = 1 \quad \text{if } \Delta\phi_s > h_{\phi_s} \\ &\text{Cflx} = 0 \quad \text{if } \Delta\phi_s \le -h_{\phi_s} \end{aligned} \tag{8}$$

where h_{ϕ_s} is hysteresis band of stator flux.

The stator flux error is defined by the difference between the reference value of flux and the actual estimated value:

$$\Delta\phi_s = |\phi_s^*| - |\phi_s| \tag{9}$$

5.2 Control of electromagnetic torque

During one sampling period, the rotor flux vector is supposed invariant. The rotor and stator flux vectors are linked by the following relation:

$$\phi_r = \frac{M}{L_s} \cdot \frac{1}{1 + j\sigma\omega T_r} \phi_s \tag{10}$$

The angle between these two vectors is given by:

$$\delta = \text{Arctan } g(\sigma\omega T_r) \tag{11}$$

Finally, between the modules of the two flux vectors, we have the following relation:

$$|\phi_r| = \frac{M}{L_s} \cdot \frac{1}{\sqrt{1 + (\sigma\omega T_r)^2}} |\phi_s| \tag{12}$$

The general expression of electromagnetic torque is given by:

$$T_{em} = p \frac{M}{\sigma L_s L_r} [\phi_s . \phi_r^*] \tag{13}$$

$$T_{em} = p \frac{M}{\sigma L_s L_r} \phi_s . \phi_r \text{Sin}(\delta) \tag{14}$$

where:

p is the number of poles pairs.

ϕ_s, ϕ_r are stator and rotor flux vectors.

δ angle between the stator and rotor flux vectors.

From expression (14), it is clear that the electromagnetic torque is controlled by the stator and rotor flux amplitudes. If those quantities are maintaining constant, the torque can be controlled by adjusting the load angle δ.

The torque regulation can be realized using three-level hysteresis comparator. (**Figure 4**).

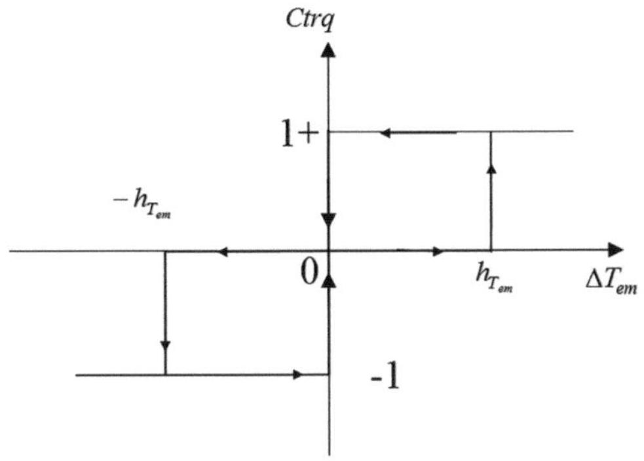

Figure 4.
Three-level hysteresis comparator for electromagnetic torque control.

The logical outputs of the torque controller are defined as:

$$\begin{array}{ll} \text{Ctrq} = 1 & \text{if } \Delta T_{\text{em}} > h_{T_{em}} \\ \text{Ctrq} = 0 & \text{if } -h_{T_{em}} \le \Delta T_{\text{em}} \le h_{T_{em}} \\ \text{Ctrq} = -1 & \text{if } \Delta T_{\text{em}} < -h_{T_{em}} \end{array} \tag{15}$$

where $h_{T_{em}}$ is hysteresis band of torque.

The torque error is defined by the difference between the references values of the torque and the actual estimated values:

$$\Delta T_{em} = T_{em}^{*} - T_{em} \tag{16}$$

6. Estimation of stator flux and electromagnetic torque

6.1 Stator flux estimation

The amplitude of the stator flux is estimated from its two-phase components $\phi_{s\alpha}$ and $\phi_{s\beta}$:

$$\phi_s = \sqrt{\phi_{s\alpha}^2 + \phi_{s\beta}^2} \tag{17}$$

Or $\phi_{s\alpha}$ and $\phi_{s\beta}$ are estimated using Eq. (4) which requires knowledge of the components of the stator current vector $i_{s\alpha}$ and $i_{s\beta}$ and that of the vector stator voltage $v_{s\alpha}$ and $v_{s\beta}$.

The stator voltage components $v_{s\alpha}$ and $v_{s\beta}$ are obtained by applying Concordia transformation on the output voltage of the three-phase VSI which are given by:

$$\begin{cases} v_{s\alpha} = \sqrt{\dfrac{2}{3}} V_{dc} \left[S_a - \dfrac{1}{2}(S_b + S_c) \right] \\ v_{s\beta} = \sqrt{\dfrac{1}{2}} V_{dc} (S_b - S_c) \end{cases} \tag{18}$$

The stator currents components $i_{s\alpha}$ and $i_{s\beta}$ can be obtained also by applying Concordia transformation on the measured currents:

$$\begin{cases} i_{s\alpha} = \sqrt{\frac{3}{2}} i_{sa} \\ i_{s\beta} = \frac{1}{\sqrt{2}} (i_{sb} - i_{sc}) \end{cases} \tag{19}$$

6.2 Electromagnetic torque estimation

The produced electromagnetic torque of the induction motor can be determined using the cross product of the stator quantities (i.e., stator flux and stator currents). The torque formula is expressed as the following:

$$T_{em} = p(\phi_{s\alpha} i_{s\beta} - \phi_{s\beta} i_{s\alpha}) \tag{20}$$

7. Switching table construction and control algorithm design

To maintain a decoupled control, a pair of hysteresis comparators receives the stator flux and torque errors as inputs. Then, the comparators outputs determine the appropriate voltage vector selection. However, the choice of voltage vector is not only depending on the output of hysteresis controllers but on the position of stator flux vector also. Thus, the circular stator flux vector trajectory will be divided into six symmetrical sectors (**Table 1**).

For each sector, the vectors (V_i *and* V_{3+i}) are not considered because both of them can increase or decrease the torque in the same sector according to the position of flux vector on the first or the second sector. If the zero vectors V_0 and V_7 are selected, the stator flux will stop moving, its magnitude will not change, and the electromagnetic torque will decrease, but not as much as when the active voltage vectors are selected. The resulting look-up table for DTC which was proposed by Takahashi is presented in **Table 2**.

	Increases	**Decreases**
ϕ_s	V_{i-1} *and* V_{i+1}	V_{i+2} *and* V_{i-2}
T_{em}	V_{i+1} *and* V_{i+2}	V_{i-1} *and* V_{i-2}

Table 1.
Generalized switching table.

Flux	**Torque**	**1**	**2**	**3**	**4**	**5**	**6**	**Comparator**
Cflx = 1	Ctrq = 1	V_2	V_3	V_4	V_5	V_6	V_1	Two-level
	Ctrq = 0	V_7	V_0	V_7	V_0	V_7	V_0	
	Ctrq = −1	V_6	V_1	V_2	V_3	V_4	V_5	Three-level
Cflx = 0	Ctrq = 1	V_3	V_4	V_5	V_6	V_1	V_2	Two-level
	Ctrq = 0	V_0	V_7	V_0	V_7	V_0	V_7	
	Ctrq = −1	V_5	V_6	V_1	V_2	V_3	V_4	Three-level

Table 2.
Look-up table for basic direct torque control.

8. Global scheme of conventional direct torque control

The global control scheme of conventional direct torque control strategy is shown in **Figure 5**. It is composed of speed regulation loop; the proportional-integral (PI) controller is used for the regulation. It is performed by comparing the speed reference signal to the actual measured speed value. Then the comparison error becomes the input of the PI controller. The pole placement method is used to determine the controller gains. The used PI controller in our work in the outer speed loop is the anti-windup controller. It allows to enhance speed control performance by canceling the windup phenomenon which is caused by the saturation of the pure integrator [20]. **Figure 6** shows the speed anti-windup PI controller diagram block.

This strategy consists on the correction of the integral action based on the difference between the control signal and the saturation limit. The difference value

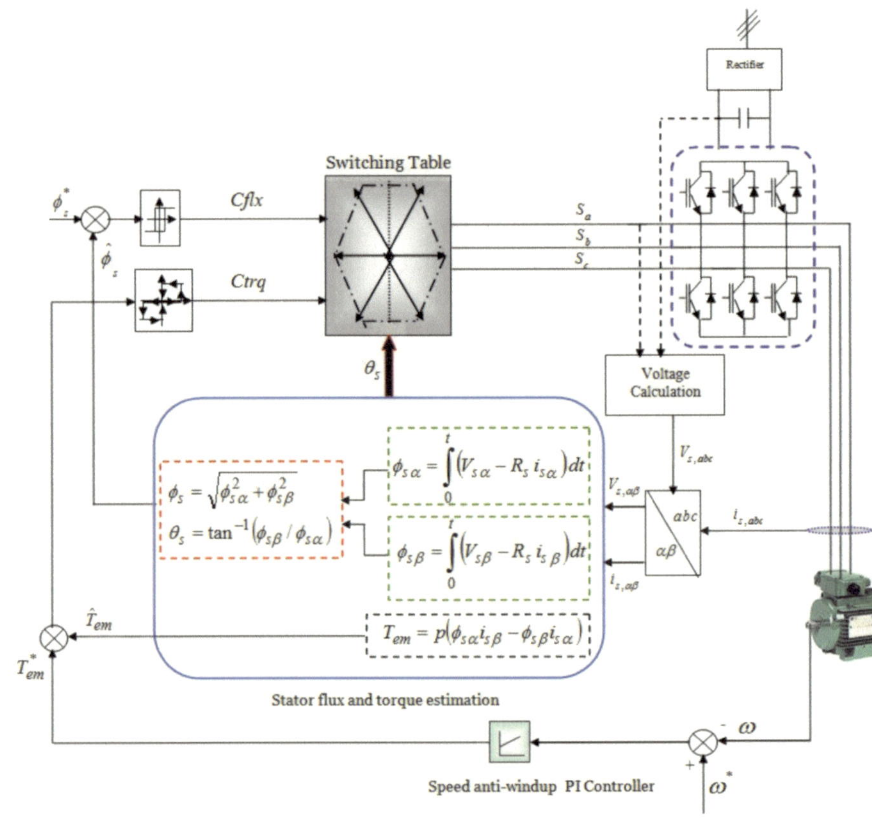

Figure 5.
Global control scheme of basic direct torque control.

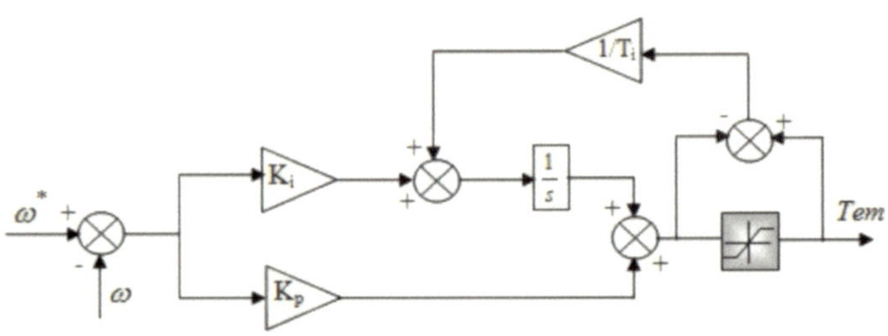

Figure 6.
Speed anti-windup PI controller.

is passed through a gain block (tracking time constant T_i) before arriving as feedback to the integrator. As well flux and torque hysteresis controllers, look-up switching table, an association of VSI-Induction motor, voltage and current calculation blocks with 3/2 (Concordia) transformation and flux/torque estimators with position/sector determination.

9. Constant switching frequency direct torque control using SVM

The conventional direct torque control has several disadvantages, among which the variable switching frequency and the high level of ripples. Consequently, they lead to high-current harmonics and an acoustical noise and they degrade the control performance especially at low speed values. The ripples are affected proportionally by the width of the hysteresis band. However, even with choosing a reduced bandwidth values, the ripples are still important due to the discrete nature of the hysteresis controllers. Moreover, the very small values of bandwidths increase inverter switching frequency. In order to overcome these drawbacks, most of the studies presented in the literature have been oriented towards modification in the conventional DTC method by the introduction of a vector modulator [21, 22]. The vector PWM technique (SVM) is used to apply a voltage vector with a fixed switching frequency. The control system consists of replacing the switching table and the hysteresis comparators with proportional and integrating controllers (PI) for controlling the stator flux and the electromagnetic torque, [6, 23–27]. The main drawbacks of DTC-SVM using PI controllers are the sensitivity of the performances to the system-parameter variations and the inadequate rejection of external disturbances and load changes [28, 29]. To cope with this disadvantage, it is suggested to replace the conventional regulators used for the speed control, flux, and electromagnetic torque by intelligent controllers by adaptive fuzzy-PI and fuzzy logic to make the control more robust against the disturbances of the parameters of the machine.

9.1 Space vector modulation algorithm

This technique is much requested in the field of control in that the reference voltages are given by a global control vector approximated over a modulation period T_z. The principle of SVM is the prediction of inverter voltage vector by the projection of the reference vector V_s^* between adjacent vectors corresponding to two non-zero switching states. For two-level inverters, the switching vector diagram forms a hexagon divided into six sectors, each one is expanded by 60° as shown in **Figure 7**.

The application time for each vector can be obtained by vector calculations, and the rest of the time period will be spent by applying the null vector.

When the reference voltage is in sector 1 (**Figure 8**), it can be synthesized by using the vectors V_1, V_2, and V_0 (zero vector).

The determination of times T1 and T2 corresponding to voltage vectors are obtained by simple projections (**Figure 9**).

$$T_1 = \frac{\sqrt{6} \cdot V_{s\alpha} - \sqrt{2} \cdot V_{s\beta}}{2V_{dc}} T \tag{21}$$

$$T_2 = \frac{\sqrt{2} \cdot V_{s\beta}}{V_{dc}} \cdot T \tag{22}$$

where V_{dc} is the DC bus voltage.

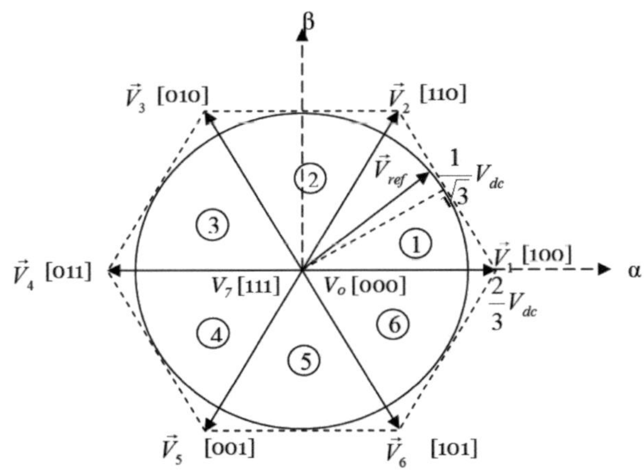

Figure 7.
Diagram of voltage space vector.

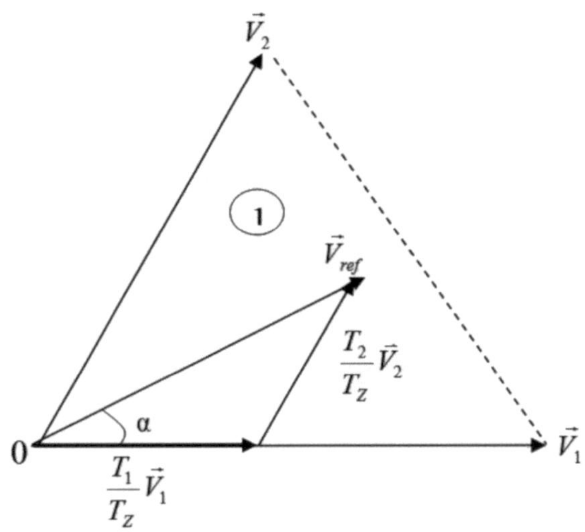

Figure 8.
Reference vector as a combination of adjacent vectors at sector 1.

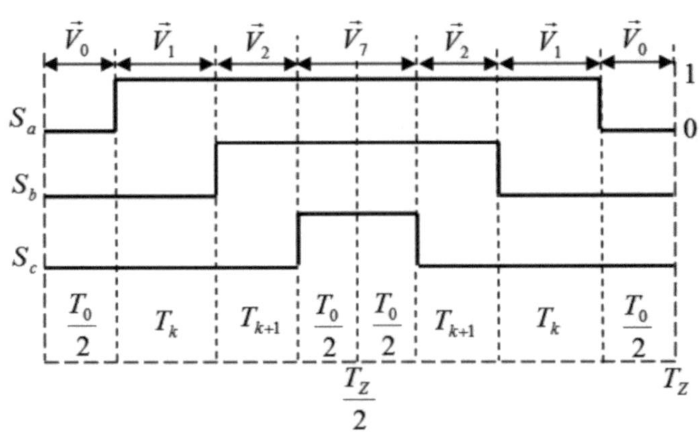

Figure 9.
Switching times of sector 1.

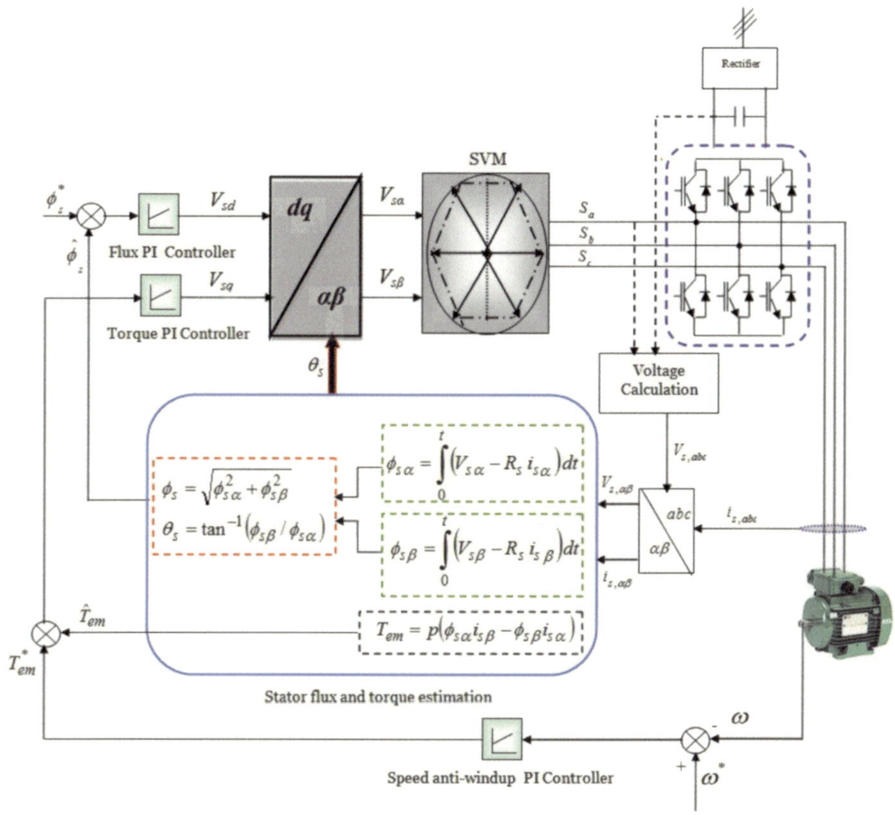

Figure 10.
Global control scheme of SVM-direct torque control with PI controller.

T_1, T_2, and T_0 are the corresponding application times of the voltage vectors, respectively. T_z is the sampling time.

Figure 10 shows the global block diagram of DTC with SVM.

9.2 Principle of fuzzy direct torque control

The complete block diagram DTC-SVM improvement of induction motor drive with fuzzy logic controller is shown in **Figure 11**. The practical difficulty with PI controllers has been addressed in the previous section. The PI controllers are being replaced by fuzzy logic controllers that generates the module and the voltage vector angle in order to bring the stator flux and the electromagnetic torque to references optimally; this vector is used by a PWM control vector to generate the pulses for the control of the switches of the inverter, and PI speed controller is replaced by the adaptive fuzzy-PI speed controller to offer a good insensitivity to parameter variations, to get better response in external disturbance rejection and fast dynamics.

9.2.1 Selecting the position of the voltage vector

The position of the reference voltage vector with respect to the stator flux vector must be chosen so as to maintain the stator flux and the electromagnetic torque in an optimal error band around their reference value. The errors of torque and flux are multiplied by "scales factors" to obtain standardized sizes and functions. These

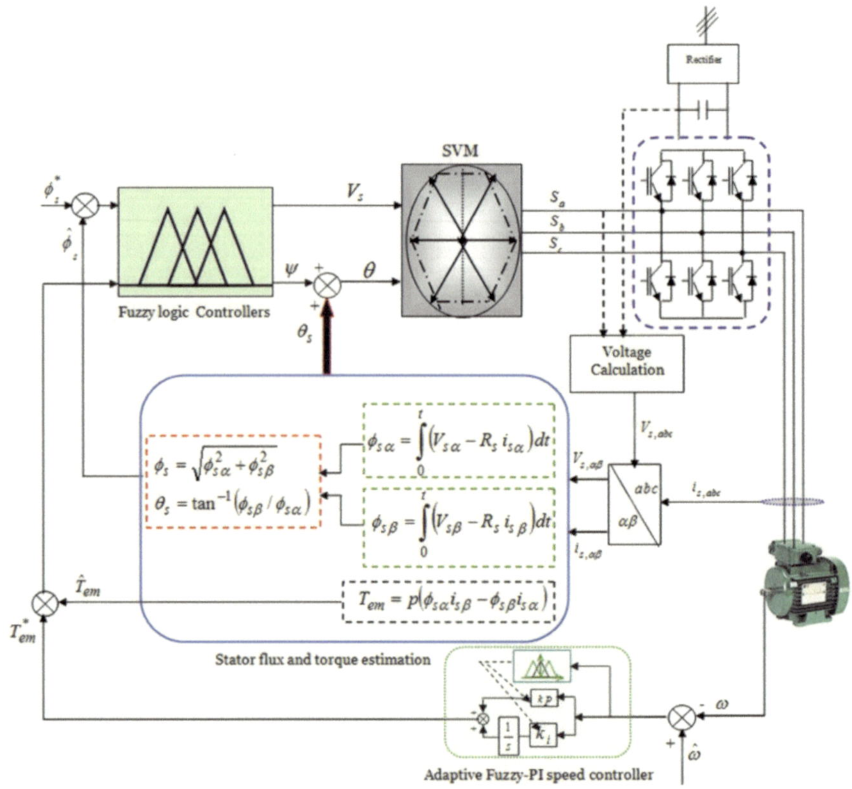

Figure 11.
Global control scheme of SVM-direct torque control with fuzzy logic controllers and adaptive fuzzy-PI speed controller.

values are used by the fuzzification block to be transformed into fuzzy values. These are used by the block fuzzy control rules after defuzzification; the value of (ψ) which must be added to the angle of the stator flux [30–32] (**Figure 12**).

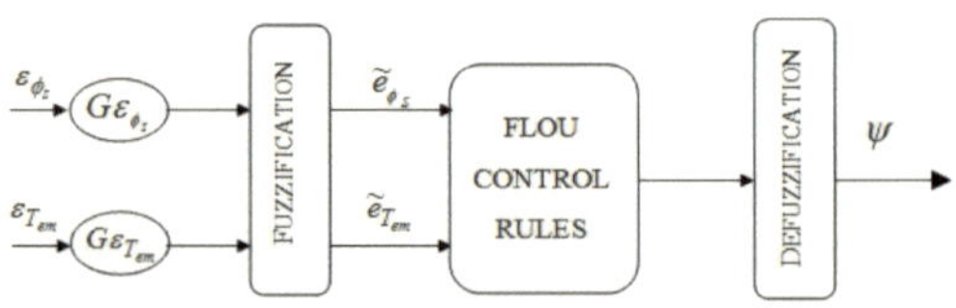

Figure 12.
Controller structure for estimating the angle (ψ).

9.2.2 Selection of the voltage vector magnitude

The voltage vector module must be selected to minimize the error of torque and flux. A fuzzy logic controller is designed to generate the appropriate voltage vector magnitude (**Figure 13**).

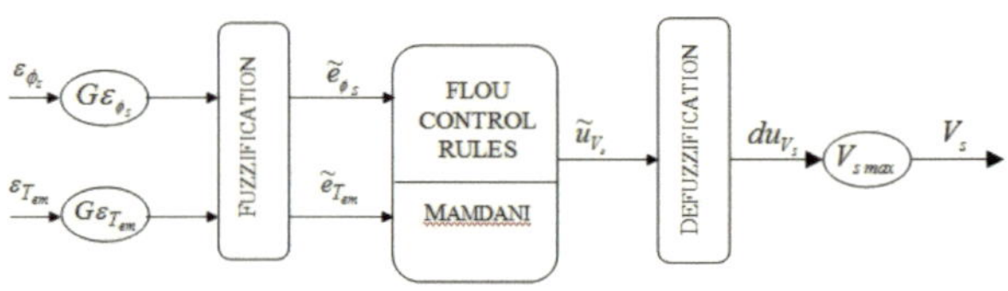

Figure 13.
Controller structure for voltage vector module estimation.

9.2.3 Selection of the voltage vector magnitude

The voltage vector obtained from the characteristic comes to the vector modulation $|\overline{V}_s| = f(\varepsilon_{T_{em}})$ which in turn generates the states, and switches using V_α, V_β the following algorithm:

Calculate the biphasic components of the desired voltage vector using the following equations [30–32]:

$$\begin{cases} V_{s\alpha} = V_s \cos(\theta) \\ V_{s\beta} = V_s \sin(\theta) \end{cases} \tag{23}$$

- Calculation of the area where the desired voltage vector is.
- Get the switching vectors and their operating cycle. Then calculate the operating cycle of the null switching vector $(t_0 = 1 - t_1 - t_2)$.
- Calculation of the relative position of the clock (PRH) in the sampling time by using the following equations:

$$\text{PRH} = \text{Rem}(\text{t}/\text{Ts})/\text{Ts}. \tag{24}$$

9.2.4 The speed control of the IM by an adaptive fuzzy-PI controller

In what follows, we show the synthesis and description of the adaptation of the PI controller by a fuzzy system method:

The fuzzy inference mechanism adjusts the PI parameters and generates new parameters during the process control. It enlarges the operating area of the linear controller (PI) so that it also works with a nonlinear system [33, 34].

The inputs of the fuzzy adapter are the error ($\boldsymbol{e}$) and the derivative of error (Δe); the outputs are the normalized value of the proportional action (k_p') and the normalized value of the integral action (k_i').

The normalization PI parameters are given by:

$$k_p' = (k_p - k_{p\,\min})/(k_{p\,\max} - k_{p\,\min}) \tag{25}$$

$$k_i' = (k_i - k_{i\,\min})/(k_{i\,\max} - k_{i\,\min}) \tag{26}$$

The parameters k_p' and k_i' are determined by a set of fuzzy rules of the form:

$$\text{If } \boldsymbol{e} \text{ is } \boldsymbol{A_i} \text{ and } \Delta e \text{ is } \boldsymbol{B_i}\text{, then } k_p' \text{ is } \boldsymbol{C_i}\text{, and } k_i' \text{ is } \boldsymbol{D_i} \tag{27}$$

where $\boldsymbol{A_i}$, $\boldsymbol{B_i}$, $\boldsymbol{C_i}$, and $\boldsymbol{D_i}$ are fuzzy sets on corresponding supporting sets.

The associated fuzzy sets involved in the fuzzy control rules are defined as follows:

PB	Positive big	**NB**	Negative big	**B**	Big
PM	Positive medium	**NM**	Negative medium	**ZE**	Zero
PS	Positive small	**NS**	Negative small	**S**	Small

The membership functions for the fuzzy sets corresponding to the error e and Δe and the adjusted proportional and integral terms (k_p' and k_i') are defined in **Figures 14** and **15**.

By using the membership functions shown in **Figure 15**, we satisfy the following condition.

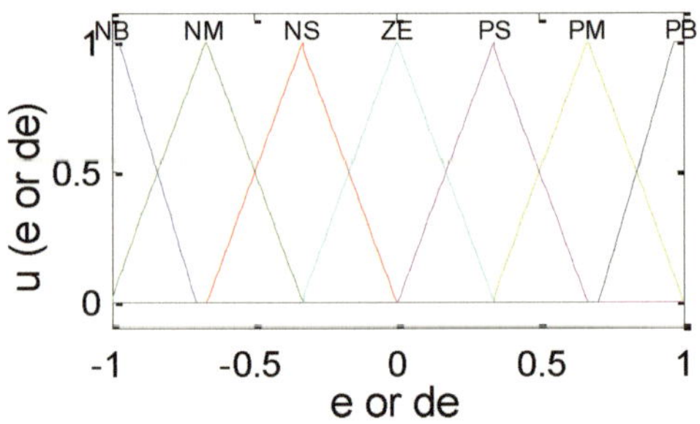

Figure 14.
Membership functions e and Δe.

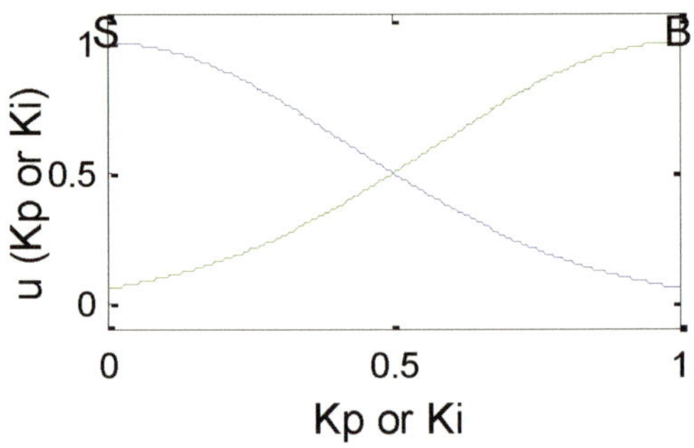

Figure 15.
Membership functions k_p' and k_i'.

$$\sum_{1}^{m} v_i = 1 \tag{28}$$

The fuzzy outputs k_p' and k_i' can be calculated by the center of area defuzzification as:

$$\left[k_p', k_i'\right] = \frac{\sum_{i=1}^{3} w_i c_i}{\sum_{i=1}^{3} w_i} = \frac{\left[c_1 \ \dots \ c_2\right]\begin{bmatrix} w_1 \\ \vdots \\ w_2 \end{bmatrix}}{\sum_{i=1}^{2} w_i} = v^T W \tag{29}$$

	e						
Δe	**NB**	**NM**	**NS**	**ZE**	**PS**	**PM**	**PB**
NB	B	B	B	B	B	B	B
NM	S	B	B	B	B	B	S
NS	S	S	B	B	B	S	S
ZE	S	S	S	B	S	S	S
PS	S	S	B	B	B	S	S
PM	S	B	B	B	B	B	S
PB	B	B	B	B	B	B	B

Table 3.
Fuzzy rule base for computing k_p'.

	e						
Δe	**NB**	**NM**	**NS**	**ZE**	**PS**	**PM**	**PB**
NB	B	B	B	B	B	B	B
NM	B	S	S	S	S	S	B
NS	B	B	S	S	S	B	B
ZE	B	B	B	S	B	B	B
PS	B	B	S	S	S	B	B
PM	B	S	S	S	S	S	B
PB	B	B	B	B	B	B	B

Table 4.
Fuzzy rules base for computing k_i'.

where $C = [c_1 \ldots c_2]$ is the vector containing the output fuzzy centers of the membership functions, $W = [w_1 \cdots w_2]/\sum_{i=1}^{2} w_i$ is the firing strength vector,

and v_i represents the membership value of the output k_p' or k_i' to output fuzzy set i.

Once the values of k_p' and k_i' are obtained (**Tables 3** and **4**), the new parameters of PI controller is calculated by the following equations:

$$k_p = (k_{p\,\max} - k_{p\,\min})k_p' + k_{p\,\min} \tag{30}$$

$$k_i = (k_{i\,\max} - k_{i\,\min})k_i' + k_{i\,\min} \tag{31}$$

10. Simulation results

The DTC control algorithms have been simulated by MATLAB/Simulink software. A comparative study between the three strategies for the direct torque control (DTC) of induction motor (IM) is presented. The first method is a conventional direct torque control (C-DTC) where the torque and the flux are regulated by the hysteresis controllers. The second one is direct torque control by space vector modulation strategy (SVM-DTC) where the torque and flux are regulated by PI controllers. The third one is fuzzy SVM-DTC with adaptive fuzzy-PI speed controller where the torque and flux are regulated by fuzzy logic controllers is presented. The simulation has been conducted for a three-phase 1.5 kW squirrel-cage induction motor with characteristics given in the appendix. The starting up and the steady states of the controlled motor with load introduction are presented. For the classical DTC, the chosen bandwidths of the hysteresis controllers are ±0.01 Wb for flux and ±0.1 Nm for torque.

This section presents the starting up state of the induction motor according to speed step reference of 1000 rpm. Then, a load of 10 Nm is suddenly applied between (t = 1 s) and (t = 2 s).

Figures 16 and **17** show, respectively, rotor speed, torque, stator phase current i_{sa}, flux magnitude, and the circular trajectory.

Figures 16 and **17(a)** illustrate the comparison between speed responses of conventional DTC and SVM-DTC-PI, according to the speed reference step of 1000 rpm. The load disturbance has been introduced between (t = 1 s) and (t = 2 s). The results of **Figure 16** show that the conventional DTC technique gives a good dynamic at starting up. We can notice that the speed regulation loop rejects the applied load disturbance quickly. The SVM-DTC-PI in **Figure 17(a)** kept the same fast speed response of DTC strategy. Since the same PI speed controller is used for both schemes, there is no difference in the transient response.

Then, the results illustrate the torque responses with load application. The figures show that at the beginning the speed controller (PI anti-windup) operates the system at the physical limit. It can be seen clearly that the constant switching frequency-based DTC strategy in **Figure 17(a)** has a reducer ripples level owing to the use of SVM compared to the conventional DTC in **Figure 16**, where it is observed that the high torque ripples exceed the hysteresis boundary. Next, the

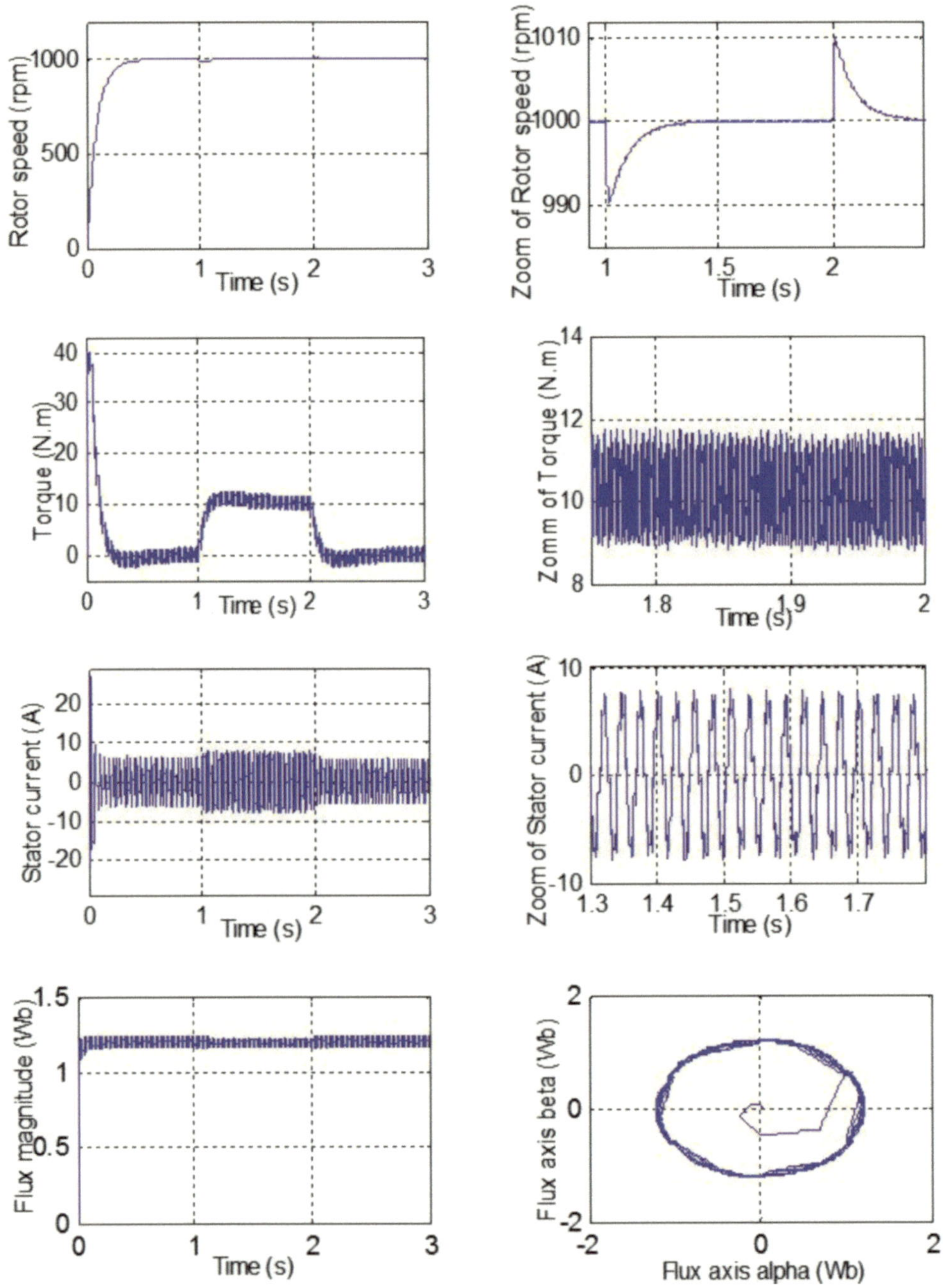

Figure 16.
Simulation results of the classical DTC control applied to IM.

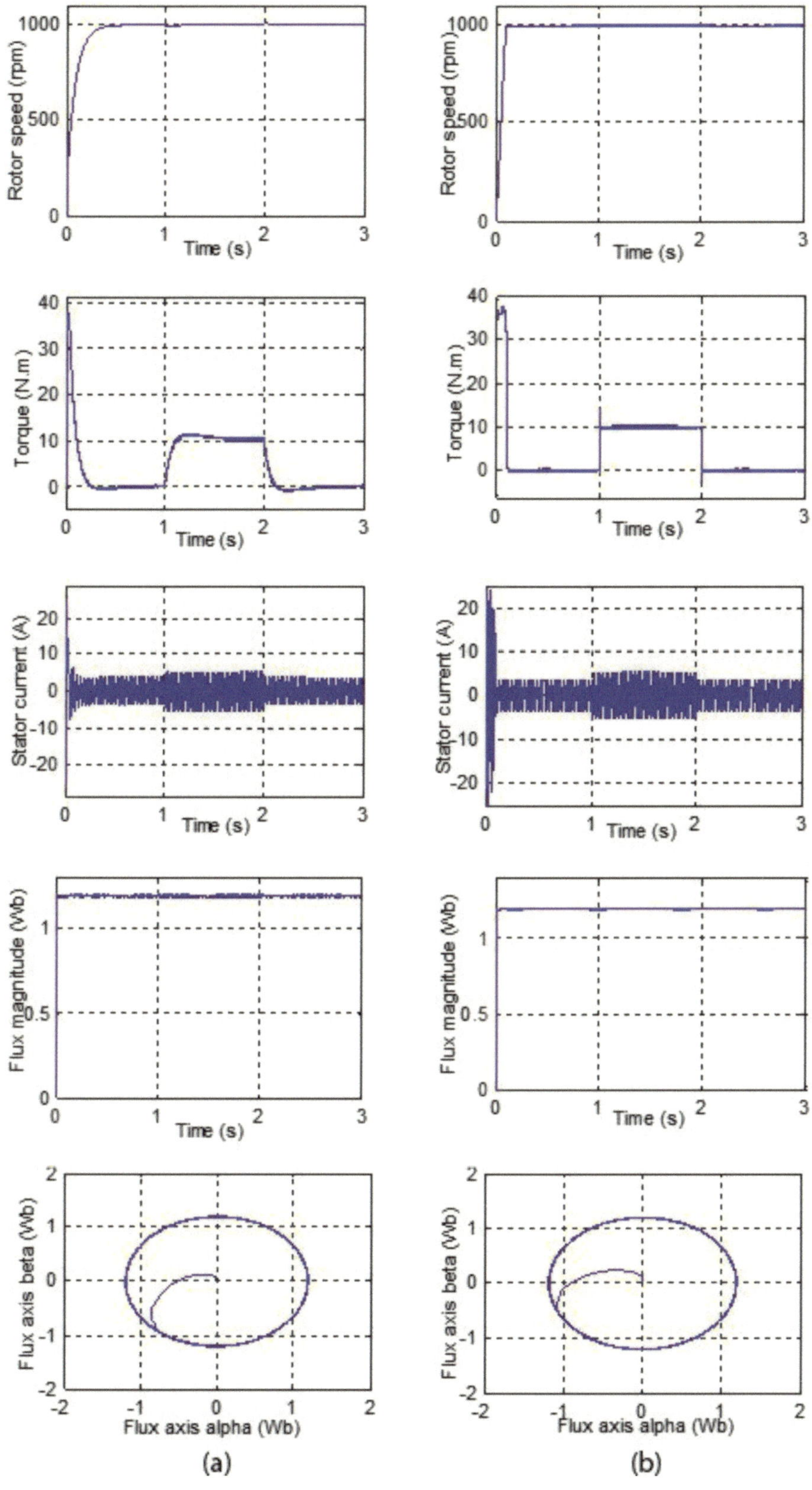

Figure 17.
Simulation results of the SVM-DTC control applied to IM. (a) SVM-DTC-PI. (b) SVM-DTC- Fuzzy.

stator phase current with zoom is presented. The conventional DTC in **Figure 16** shows a chopped sinusoid waveform of current which indicates a high harmonic level, while SVM-DTC in **Figure 17(a)** shows a smoother sinusoid waveform. After that, the results exhibit the magnitude of stator flux evolution and circular trajectory. It is clear that the flux ripples of the conventional DTC have exceeded the hysteresis boundary. The magnitude and the trajectory illustrate that the flux takes a few steps before reaching the reference value (1.2 Wb) at the starting stage due to the zone's changing.

The simulation in **Figure 17(b)** shows that the SVM-DTC-fuzzy has better performance than those obtained by both other DTC strategies (conventional and SVM-PI). There is an appreciable decrease in the start-up response time; we can notice that the speed regulation loop rejects the applied load disturbance very quickly which proves the performance of adaptive fuzzy-PI controller as well as a significant attenuation of the ripples of the torque and of the sinusoidal current without any ripple in the steady state.

11. Conclusion

The main objective of this chapter is the improvement of the performance of an induction motor drive controlled by DTC. The objective of this improvement is to minimize the ripples of the couple and the flux of the IM on the one hand and the decrease of the switching frequency of the inverter on the other hand. In this context, a comparative analysis between different DTC strategies has been presented. This chapter began by explaining the principle of the conventional DTC, SVM-DTC-PI, and SVM-DTC-fuzzy with adaptive Fuzzy-PI speed controller. The chapter presents later a discussion based on the simulation results presented in the same work. The synthesis of this simulation study reveals advantages of SVM-DTC-fuzzy scheme compared to the two strategies: conventional DTC and SVM-DTC-PI. It has been observed by comparing the torque, speed, and stator flux characteristics that the method SVM-DTC-fuzzy is better. It is clear that the current is sinusoidal without any ripple in the steady state and torque ripples are reduced. In order to improve the SVM-DTC-fuzzy to have better performances, this method has been associated to the adaptive fuzzy-PI speed controller. This association makes the induction motor-based DTC perform more and more stable; there is an appreciable decrease in the start-up response time; we can notice that the speed regulation loop rejects the applied load disturbance very quickly.

A. IM motor parameters

Item	Symbol	Data
IM mechanical power	P_W	1.5 kw
Nominal speed	ω	1420 rpm
Nominal frequency	f	50 Hz
Pole pair number	P	2
Stator resistance	R_s	4.85 Ω
Rotor resistance	R_r	3.805 Ω
Stator self-inductance	L_s	274 mH

Item	Symbol	Data
Rotor self-inductance	L_r	274 mH
Mutual inductance	L_m	258 mH
Moment of inertia	J	0.031 kg m^2
Friction coefficient	F	0.00114 kg m^2/s

Author details

Cherifi Djamila* and Miloud Yahia
GACA Laboratory, University of Dr Moulay Tahar, Saida, Algeria

*Address all correspondence to: d_cherifi@yahoo.fr

IntechOpen

© 2020 The Author(s). Licensee IntechOpen. This chapter is distributed under the terms of the Creative Commons Attribution License (http://creativecommons.org/licenses/by/3.0), which permits unrestricted use, distribution, and reproduction in any medium, provided the original work is properly cited.

References

[1] Takahashi, Noguchi T. A new quick-response and high-efficiency control strategy of an induction motor. IEEE Transactions on Industry Applications., Vol.22, No.5, pp.820–827, 1986. DOI: 10.1109/TIA.1986.4504799

[2] Depenbrock M. Direct self control (DSC) of inverter fed induction machine. IEEE Transactions on Power Electronics. 1988;**3**(4):420-429. DOI: 10.1109/63.17963

[3] Ammar A, Benakcha A, Bourek A. Closed loop torque SVM-DTC based on robust super twisting speed controller for induction motor drive with efficiency optimization. International Journal of Hydrogen Energy. 2017:1-13. DOI: 10.1016/j.ijhydene.2017.04.034

[4] Lazim MT, Al-Khishali MJM, Al-Shawi AI. Space vector modulation direct torque speed control of induction motor. Procedia Computer Science. 2011;**5**:505-512. DOI: 10.1016/j.procs.2011.07.065

[5] Habetler Thomas G, Profumo F, Pastorelli M, Tolbert Leon M. Direct torque control of induction machines using space vector modulation. IEEE Transactions on Industry Applications. 1992;**28**(5):1045-1053. DOI: 10.1109/28.158828

[6] Hassan Farhan R, Nadia Tan ML, Koh SP, Abdalla AN, Chong KH, Tiong SK. DTC-SVM based on PI torque and PI flux controllers to achieve high performance of induction motor. Research Journal of Applied Sciences, Engineering and Technology. 2014;**7**(4):875-891. DOI: 10.19026/rjaset.7.330

[7] Krim S, Gdaim S, Mtibaa A, Mimoun MF. FPGA-based DTC-SVM: A comparative study between two approaches. IEEE. 2016. DOI: 10.1109/CEIT.2016.7929060

[8] Yu H, Chen Z. Three-phase induction motor DTC-SVPWM scheme with self-tuning PI-type fuzzy controller. International Journal of Computer and Communication Engineering. 2015;**4**(3). DOI: 10.17706/IJCCE.2015.4.3.204-210

[9] Ben Salem F, Derbel N. Direct torque control of induction motors based on discrete space vector modulation using adaptive sliding mode control. International Journal of Electric Power Components and Systems. 2014;**42**(14):1598-1610. DOI: 10.1080/15325008.2014.927029

[10] Ben Salem F, Derbel N. Performance analysis of DTC-SVM sliding mode controllers-based parameters estimator of electric motor speed drive. Mathematical Problems in Engineering. 2014:1-11. DOI: 10.1155/2014/127128. Article ID 127128

[11] Ben Salem F, Derbel N. Chapter 14: DTC-SVM based sliding mode controllers with load torque estimators for induction motor drives. In: Applications of Sliding Mode Control, Studies in Systems, Decision and Control 79. Singapore: Springer Science+Business Media; 2017. pp. 269-297. DOI: 10.1007/978-981-10-2374-3_14

[12] Brandstetter P, Chlebis P, Palacky P. Direct torque control of induction motor with direct calculation of voltage vector. Advances in Electrical and Computer Engineering Journal. 2010;**10**(4):17-22. DOI: 10.4316/aece.2010.04003

[13] Carmelia S, Maurib M. Direct torque control as variable structure control: Existence conditions verification and analysis. Electric Power Systems Research. 2011;**81**:1188-1196. DOI: 10.1016/j.epsr.2011.01.007

[14] Kuo-Kai S, Juu-Kuh L, Van-Truong P, Ming-Ji Y, Te-Wei W. Global minimum torque ripple design for direct torque control of induction motor drives. IEEE Transactions on Industrial Electronics. 2010;**57**(9):3148-3156. DOI: 10.1109/TIE.2009.2038401

[15] Nash JN. Direct torque control, induction motor vector control without an encoder. IEEE Transactions on Industry Applications. 1997;**33**(2): 333-341. DOI: 10.1109/28.567792

[16] Tiitinen P, Pohkalainen P, Lalu J. The next generation motor control method: Direct torque control (DTC). European Power Electronics and Drives. 1995;**5**(1):14-18. DOI: 10.1080/09398368.1995.11463357

[17] Kazmierkowski MP, Kasprowicz A. Improved direct torque and flux vector control of PWM inverter-fed induction motor drives. IEEE Transactions on Industrial Electronics. 1995;**45**(4): 344-350. DOI: 10.1109/41.402472

[18] Buja GS, Kazmierkowski MP. Direct torque control of PWM inverter-fed AC motors—A survey. IEEE Transactions on Industrial Electronics. 2004;**51**(4): 744-757. DOI: 10.1109/TIE.2004.831717

[19] Yuttana K, Suttichai P, Hamid AT. Modified direct torque control method for induction motor drives based on amplitude and angle control of stator flux. Journal of Electric Power Systems Research. 2008;**78**(10):1712-1718. DOI: 10.1016/j.epsr.2008.02.015

[20] Yang M, Tang S, Xu D. Comments on Antiwindup strategy for PI-type speed controller. IEEE Transactions on Industrial Electronics. 2015;**62**(2): 1329-1332. DOI: 10.1109/TIE.2014.2363626

[21] Jose L, Azcue P, Ernesto R. Three-phase induction motor DTC-SVM scheme with self-tuning PI-type fuzzy controller. In: International Conference on FSKD. 2010. DOI: 10.1109/FSKD.2010.5569358

[22] Lascu C, Trzynadlowski AM. Combining the principles of sliding mode, direct torque control, and space-vector modulation in a high-performance sensorless AC drive. IEEE Transactions on Industry Applications. 2004;**40**(1):170-177. DOI: 10.1109/TIA.2003.821667

[23] Kyo-Beum L, Blaabjerg F. An improved DTC-SVM method for sensorless matrix converter drives using an over modulation strategy and a simple nonlinearity compensation. IEEE Transactions on Industrial Electronics. 2007;**54**(6):3155-3166. DOI: 10.1109/TIE.2007.905914

[24] Bounadja M, Belarbi A, Belmadani B. A high-performance space vector modulation—Direct torque controlled induction machine drive based on stator flux orientation technique. Advances in Electrical and Computer Engineering. 2009;**9**(2): 28-33. DOI: 10.4316/aece.2009.02005

[25] Bo L, Gui-Fang G, Xiao-Hong H, Xiao-Ning L. Novel torque ripple minimization algorithm for direct torque control of induction motor drive. Journal of Shanghai University. 2009; **13**(2):155-163. DOI: 10.1007/s11741-009-0213-3

[26] Rashag HF, Koh SP, Chong KH, Tiong SK, Tan NML, Abdalla AN. High performance of space vector modulation direct torque control SVM-DTC based on amplitude voltage and stator flux angle. Research Journal of Applied Sciences, Engineering and Technology. 2013;**5**(15):3934-3940. DOI: 10.19026/rjaset.5.4455

[27] Subodh KB, Kiran KJ. Five-phase induction motor DTC-SVM scheme with PI controller and ANN controller. Procedia Technology. 2016:816-823. DOI: 10.1016/j.protcy.2016.08.184

[28] Gadoue SM, Giaouris D, Finch JW. Artificial intelligence-based speed control of DTC induction motor drives —A comparative study. Electric Power Systems Research. 2009:210-219. DOI: 10.1016/j.epsr.2008.05.024

[29] El Ouanjli N, Derouich A, El Ghzizal A, Motahhir S, Chebabhi A, El Mourabit Y, et al. Modern improvement techniques of direct torque control for induction motor drives—A review. Protection and Control of Modern Power Systems. 2019:1-12. DOI: 10.1186/s41601-019-0125-5

[30] Uddin M, Hafeez M. FLC-based DTC scheme to improve the dynamic performance of an IM drive. IEEE Transactions on Industry Applications. 2012;**48**(2):823-831. DOI: 10.1109/TIA.2011.2181287

[31] Ouledali O, Meroufel A, Wira P, Bentouba S. Direct torque fuzzy control of PMSM based on SVM. Energy Procedia. 2015:1314-1322. DOI: 10.1016/j.egypro.2015.07.778

[32] Nour MO, Tedjini HA. Improved fuzzy logic control strategy of induction machine based on direct torque control. International Journal of Mechatronics, Electrical and Computer Technology (IJMEC). DOI: 649123/10.225134

[33] Bousserhane IK, Hazzab A, Rahli M, Kamli M, Mazari B. Adaptive PI controller using fuzzy system optimized by genetic algorithm for induction motor control. IEEE. 2006:1-8. DOI: 10.1109/CIEP.2006.312162

[34] Laroussi K, Zelmat M. Implementation of a fuzzy logic system to tune a PI controller applied to induction motor. Advances in Electrical and Computer Engineering. 2009;**9**(3): 107-113. DOI: 10.4316/aece.2009.03019

Chapter 3

DTC-SVM Approaches of an Induction Motor Dedicated to Position Control Applications

Fatma Ben Salem

Abstract

The chapter is devoted to the DTC and DTC-SVM position control approaches of induction motor (IM) allowing the movement of a photovoltaic panel according to the maximum sunshine position to extract a high efficiency of the system. The DTC is selected to full the application requirements, especially a maximum torque at standstill. This feature is necessary in order to guarantee a high degree of robustness of the maximum sunshine position tracking system against the high and sudden load torque variations characterized by the gusts of wind. The first step is devoted to a comparison study between three DTC strategies, dedicated to position control, such that: the basic DTC strategy, the DTC strategy with a look-up table including only active voltage vectors, and the DTC-SVM strategy with hysteresis controllers. Furthermore, the synthesis and the implementation of DTC-SVM approaches based on position control are treated. Within this context, the final part of the chapter proposes a comparison between three DTC-SVM approaches: (i) a DTC-SVM approach using PI controllers, (ii) a DTC-SVM approach using PI controllers with a nonlinear compensator, and (iii) a DTC-SVM approach using sliding mode controllers. In that case, an adaptation approach of parameter estimators are implemented in order to eliminate the effects of parameter variations and load disturbances. Simulations results show that the SM DTC-SVM approach gives the best results.

Keywords: induction machine, position control, SM, DTC, SVM, parameters variations, load disturbances

1. Introduction

Induction motors (IM) are very common because they are inexpensive and robust, finding use in everything from industrial applications such as pumps, fans, and blowers to home appliances. In recent years, the control of high-performance IM drives for general industrial applications and production areas has received a lot of research interests.

The most modern technique, for the induction machine, is the direct torque and the stator flux vector control method (DTC). It has been realized in an industrial way by ABB, using the theoretical background proposed by *Takahashi* [1] and Depenbrock [2] in the middle of 1980's. Over the years, DTC method becomes one of the high-performance control strategies for AC machines to provide a very fast torque and flux control [3, 4].

The DTC has been selected in order to fulfill the application requirements, especially a maximum torque at standstill. This feature is necessary in order to guarantee a high degree of robustness of the maximum sunshine position tracking system against the high and sudden load torque variations characterized by the gusts of wind. This positioning system can be introduced in the multi-sources hybrid system, in order to allow high efficiencies of photovoltaic systems. To do so, an electric motor drive could be associated with photovoltaic panels in order to be able to track the maximum sunshine positions during the day. In what follows, the chapter will be focused on the problem of position regulation of an induction motor under DTC and DTC-SVM strategies.

2. A case study: solar panel positioning

2.1 Problem heading

Photovoltaic panels are commonly exposed to the sun in a fixed position corresponding to the maximum sunshine recorded during a day that is the position of the sun at midday. Nevertheless, this strategy does not allow the extraction of the maximum power during a day and therefore a high efficiency of photovoltaic systems, which can be integrated with the multi-source hybrid system, described above. An approach to solve this problem consists in moving photovoltaic panels according to the maximum sunshine position. To do so, an electric motor drive could be associated with photovoltaic panels in order to be able to track the maximum sunshine positions during the day. Accounting for the high perturbation amplitude applied to the panel, the control strategy to be implemented in the drive is of great importance [5].

The proposed tracking system has two freedom degrees in such a way that it allows the displacement of the photovoltaic system within latitudes and meridians: the first degree of freedom is controlled automatically by an IM drive under the control of a DTC strategy.

The DTC approach has been selected in order to full the application requirements, especially a maximum torque at standstill. This feature is necessary in order to guarantee a high degree of robustness of the maximum sunshine position tracking system against the high and sudden load torque variations characterized by the gusts of wind.

The following work will be focused on the study of the first freedom degree. Special attention is paid to the implementation of a suitable DTC strategy in the IM drive.

2.2 Mathematical model of induction machines

The dynamic behavior of an induction machine is defined in terms of space variables in the sequel:

$$\begin{cases} \frac{d}{dt}\phi_{\alpha s} = v_{\alpha s} - R_s i_{\alpha s} \\ \frac{d}{dt}\phi_{\beta s} = v_{\beta s} - R_s i_{\beta s} \\ \frac{d}{dt}\phi_{\alpha r} = -R_r i_{\alpha r} - \omega_m \phi_{\beta r} \\ \frac{d}{dt}\phi_{\beta r} = -R_r i_{\beta r} + \omega_m \phi_{\alpha r} \end{cases} \tag{1}$$

considering that subscripts s and r refer to stator and rotor, subscripts α and β refer to components in (α, β) frame, v, i and $(\rho$ refer to voltage, current and flux, R_s and R_r refer to stator and rotor resistances, and ω_m refers to the machine speed $(\omega_m = N_p\Omega_m = \omega_s - \omega_r$ and N_p is the pole pair number).

Relationships between currents and flux are:

$$\begin{cases} \phi_{\alpha s} = L_s i_{\alpha s} + M i_{\alpha r} \\ \phi_{\alpha r} = M i_{\alpha s} + L_r i_{\alpha r} \\ \phi_{\beta r} = M i_{\beta s} + L_r i_{\beta r} \\ \phi_{\beta s} = L_s i_{\beta s} + M i_{\beta r} \end{cases} \tag{2}$$

where L and M refer to the inductance and the mutual one.

The mechanical part of the machine is described by:

$$J\frac{d}{dt}\Omega_m = T_{em} - T_l \tag{3}$$

where J is the motor inertia and T_l represent the load torque.

2.3 Voltage source inverter

The made constant DC voltage by the rectifier is delivered to the inverter input, which thanks to controlled transistor switches, converts this voltage to three-phase AC voltage signal with wide range variable voltage amplitude and frequency.

The voltage vector of the three-phase voltage inverter can be represented as follows:

$$\vec{V}_s = \sqrt{\frac{2}{3}}\left[S_a + S_b e^{j\frac{2\pi}{3}} + S_c e^{j\frac{4\pi}{3}}\right] \tag{4}$$

where S_a, S_b and S_c are three-phase inverter switching functions, which can take a logical value of either 0 or 1.

2.4 Basic concept of DTC based position control

The implementation scheme of the *Takahashi* basic DTC strategy applied to the position regulation of an induction motor drive is shown in **Figure 1**.

Referring to [5, 6], it has been found that the *Takahashi* basic DTC strategy is penalized at low speeds by the so-called "demagnetization phenomenon" which is caused by the systematic application of zero voltage vectors when the torque regulator output is zero, independently of the flux regulator output state. Indeed, the application of these voltage vectors during a sampling period T_s yields a slight decrease of the stator flux at high speeds. However, at low speeds, the application of zero voltage vectors leads to a high reduction of the stator flux, yielding the demagnetization problem which affects the electromagnetic torque.

In order to overcome this shortcoming, the zero-voltage vectors can be substituted by active ones. For a given stator flux vector and when the torque regulator output is "0", the active vector around which is located the sector including the stator flux vector, is applied. The resulting look-up table is given in **Table 1**. Nevertheless, this substitution is associated to an other crucial problem: that is an increase of the inverter switching frequency which compromises the drive efficiency.

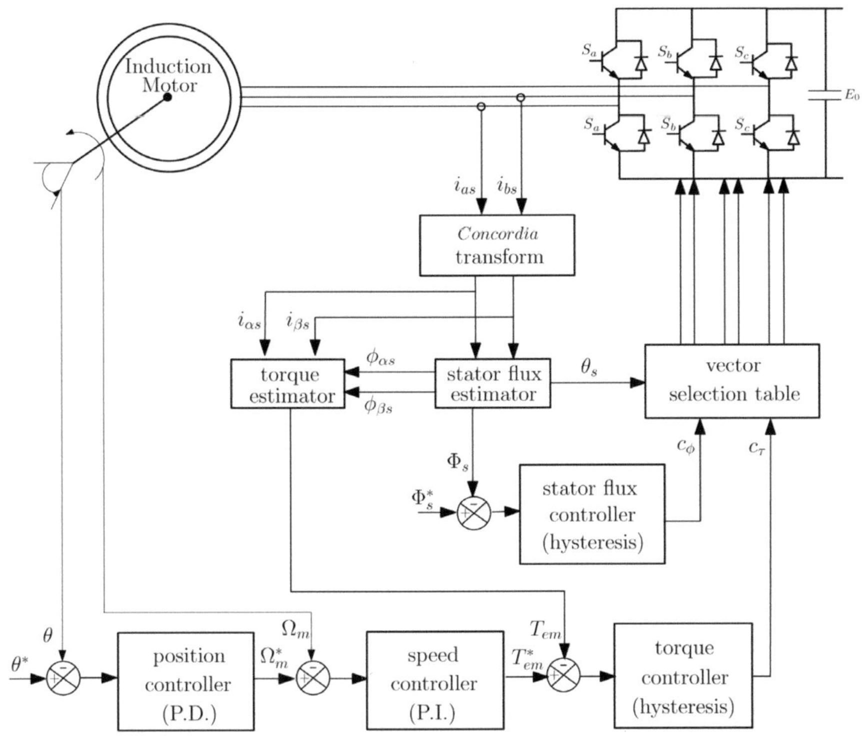

Figure 1.
IM position regulation based on basic DTC strategy.

c_ϕ		+1			−1	
c_r	+1	0	−1	+1	0	−1
S1	V_2 (100)	V_1 (101)	V_6 (001)	V_3 (110)	V_4 (010)	V_5 (011)
S2	V_3 (110)	V_2 (100)	V_1 (101)	V_4 (010)	V_5 (011)	V_6 (001)
S4	V_4 (010)	V_3 (110)	V_2 (100)	V_5 (011)	V_6 (001)	V_1 (101)
S4	V_5 (011)	V_4 (010)	V_3 (110)	V_6 (001)	V_1 (101)	V_2 (100)
S5	V_6 (001)	V_5 (011)	V_4 (010)	V_1 (101)	V_2 (100)	V_3 (110)
S6	V_1 (101)	V_6 (001)	V_5 (011)	V_2 (100)	V_3 (110)	V_4 (010)

Table 1.
Look-up table with zero-voltage vectors substituted by active ones.

2.5 Concept of DTC-SVM with hysteresis controllers based position control

The implementation scheme of the DTC-SVM strategy with hysteresis controllers applied to the position regulation of an induction motor drive is shown in **Figure 2**. It has the same layout as the one of the basic DTC strategy proposed in section II, except that the SVM bloc is added to the control system that ensures an imposed switching frequency [7–12].

DOI: http://dx.doi.org/10.5772/intechopen.94436

2.6 Simulations and discussions: A comparative study

The ratings and parameters of the induction machine, used in the simulation study, are listed in **Tables 2** and **3** respectively.

- The sampling period T_s has been chosen equal to $50\mu s$ in the cases of the first and the second strategies, whereas in the case of the third strategy, it has been chosen equal to 100μs.

- Bandwidths of flux and torque hysteresis regulators have been chosen as: $\varepsilon_\phi = 0.02\,Wb$ and $\varepsilon_\tau = 5N$ for the two first strategies, whereas for the third one, they have been selected as $\varepsilon_\phi = 0.02\,Wb$ and $\varepsilon_\tau = 3.5N$.

- The load torque is given by the following expression: $T_l = K \sin\theta$. K has been calculated and has been found equal to 57.762 N.m.

- The modulation period has been fixed to $T_{mod} = 150\mu s$ in DTC-SVM approach under study.

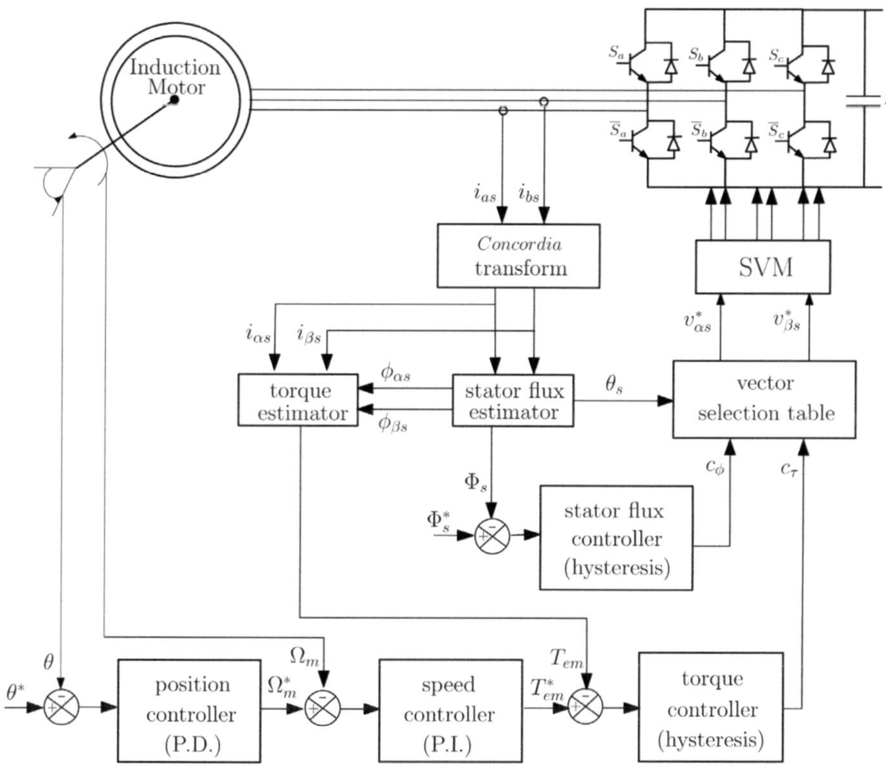

Figure 2.
Induction motor position regulation based on the DTC-SVM strategy.

Power	10 kW	Voltage	380 V/220 V
Efficiency	80%	Current	24A/41A
Speed	1500 rpm	Frequency	50 Hz

Table 2.
Induction machine ratings.

$r_s = 0.29\Omega$	$r_r = 0.38\Omega$	$M = 47.3\text{mH}$
$L_s = L_r = 50\text{mH}$	$N_p = 2$	$J = 0.5\text{Kg.m}^2$

Table 3.
Induction machine parameters.

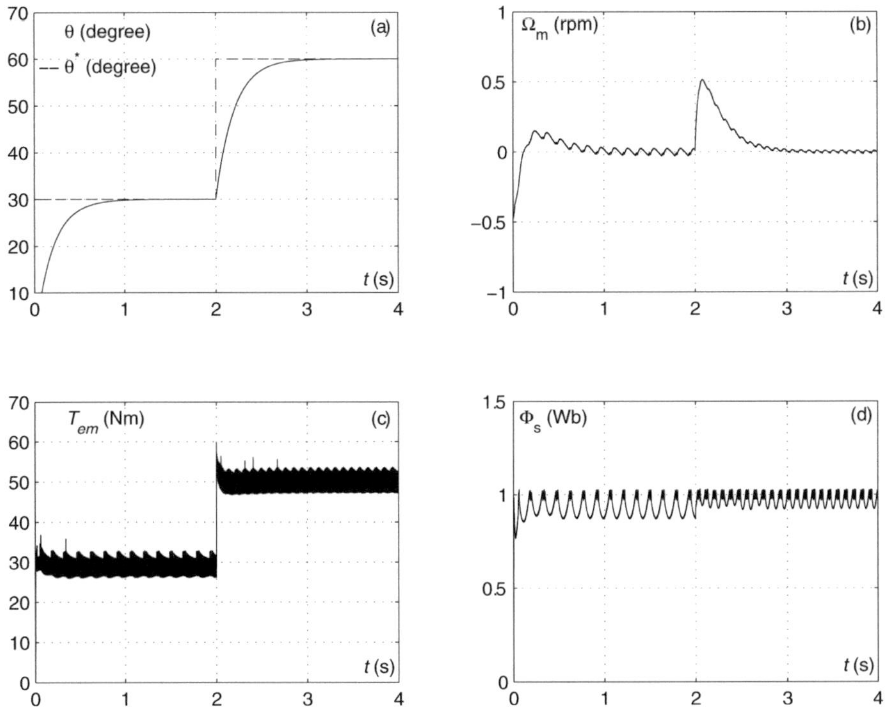

Figure 3.
Induction motor position regulation under the basic Takahashi DTC strategy, (a): Rotor position θ and its reference, (b): Speed Ω_m, (c): Electromagnetic torque T_{em}, (d): Stator flux Φ_s.

Figures 3–5 show the induction motor dynamic following the application of a dual-step reference position under the basic *Takahashi* DTC strategy, the modified *Takahashi* one, and the DTC-SVM strategy with a controlled commutation frequency, respectively. In order to highlight performances gained by the DTC-SVM scheme, resulting features are compared to the obtained ones following the implementation of the basic DTC strategy. The analysis of these results leads to the following remarks:

- **Figures 3a–5a** show that the three DTC strategies exhibit almost the same position and speed dynamics,
- Performances of the flux loop of the basic *Takahashi* DTC strategy is affected by the demagnetized phenomenon (**Figure 3d**). In fact, the analysis of the *Takahashi* strategy highlights low performances at low speed operations. Under such conditions, and for steady state operations, the motor turns to be demagnetized.
- In order to overcome the demagnetization problem caused by zero-voltage vectors included in the look-up table of the basic DTC strategy, these have

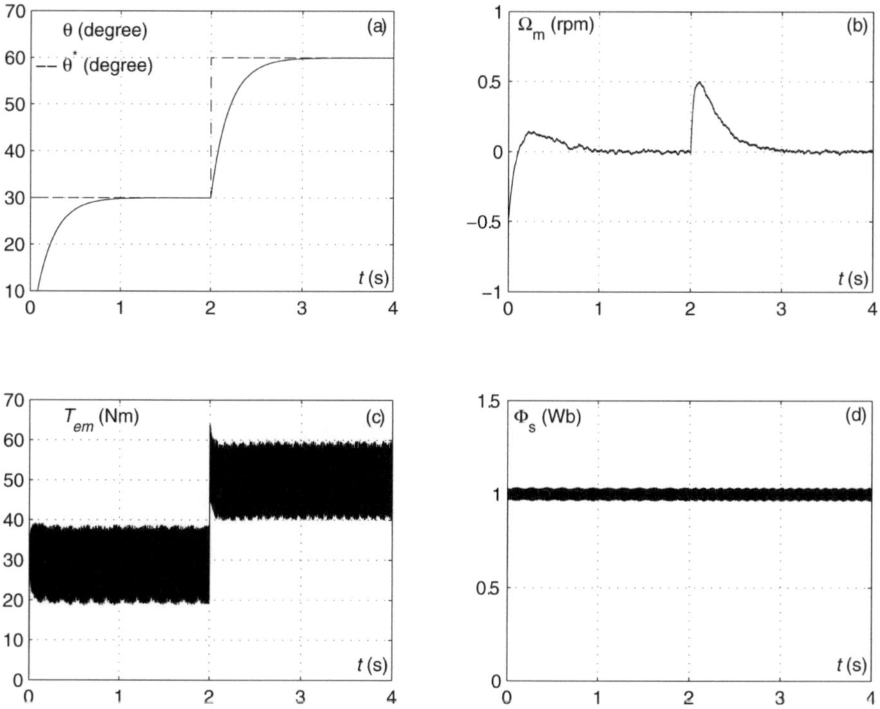

Figure 4.
Induction motor position regulation under the modified Takahashi DTC strategy, (a): Rotor position θ and its reference, (b): Speed Ω_m, (c): Electromagnetic torque T_{em}, (d): Stator flux Φ_s.

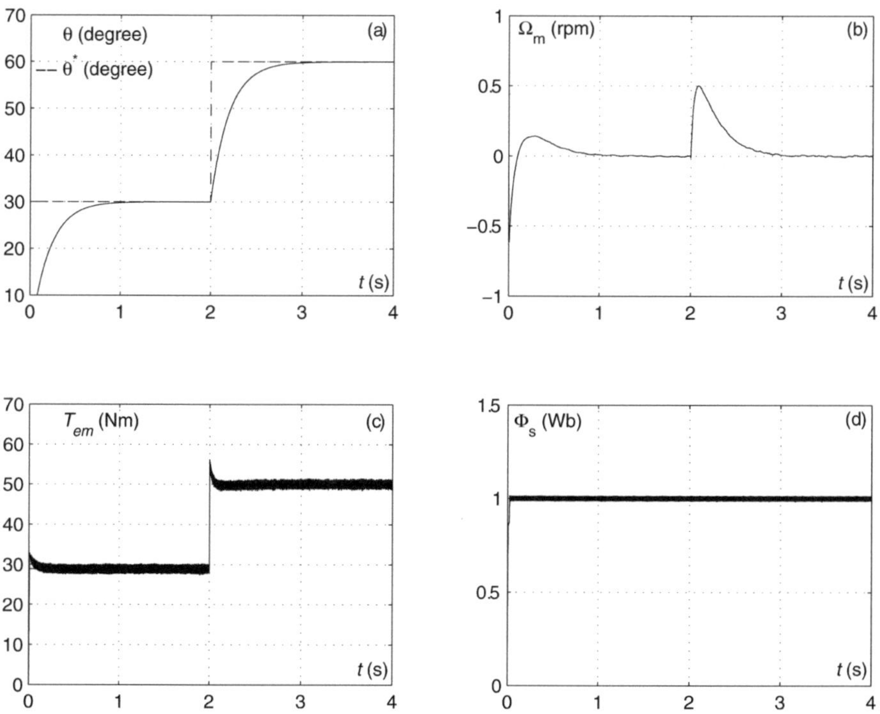

Figure 5.
Induction motor position regulation under the DTC-SVM strategy with a constant commutation frequency of 6.5 kHz, (a) rotor position θ, (b): Speed Ω_m,: (b): Electromagnetic torque T_{em}, (d): Stator flux Φ_s.

been substituted by active vectors. Obtained results are illustrated in **Figure 4**. Referring to **Figure 4d**, one can notice that the demagnetization problem has been removed, while performances of the motor, for high speeds, are not affected. However, the torque ripple amplitudes rise considerably (**Figure 4c**) with respect to the one yielded by the *Takahashi* DTC strategy, which represents a severe drawback.

- **Figure 5** shows that the DTC-SVM strategy exhibits high dynamical performances. In fact, this approach presents a low torque ripple amplitude (**Figure 5c**). Moreover, it completely eliminates the demagnetization phenomenon (**Figure 5d**).

Further investigation of the stator flux has been achieved through the representation of the stator flux vector extremity locus in the $(\alpha, \ \beta)$ plane. This has been

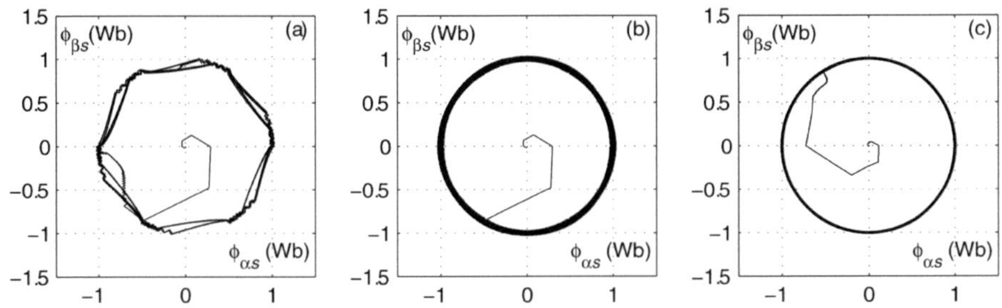

Figure 6.
Locus of the extremities of $\overline{\Phi}_s$, with (a) basic Takahashi DTC strategy, (b) modified Takahashi DTC strategy and (c) DTC-SVM strategy with a constant commutation frequency.

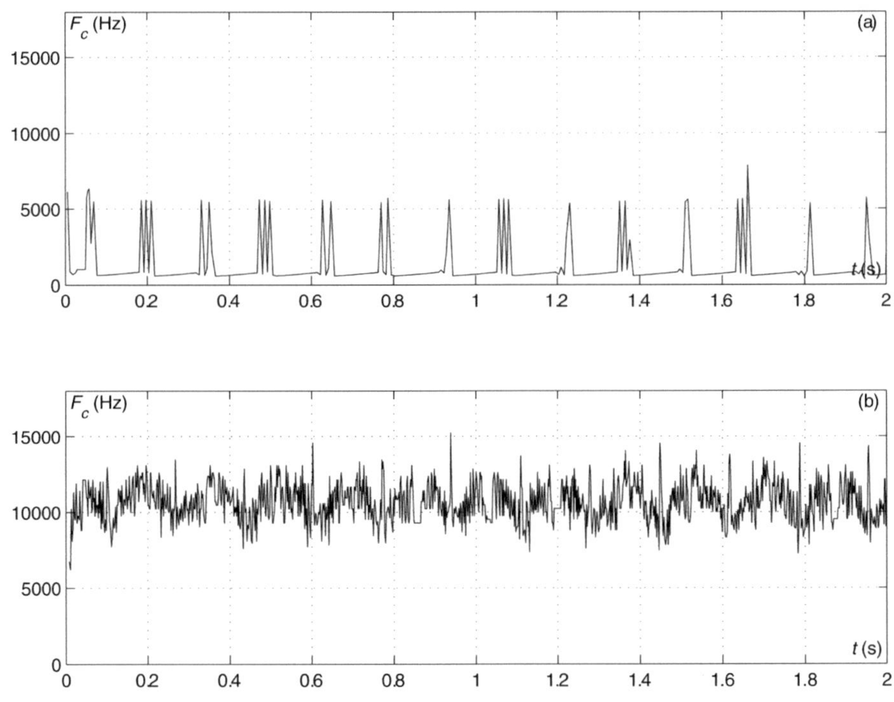

Figure 7.
Average commutation frequency of the inverter power switches, (a) basic Takahashi DTC strategy, (b) modified Takahashi DTC strategy.

done considering the three DTC strategies. Obtained results are shown in **Figure 6**. One can notice, easily, that the DTC-SVM strategy with hysteresis regulators and with an imposed commutation frequency yields to the smoothest circular locus.

Finally, we have involved in the assessment of the average commutation frequencies of both basic and modified *Takahashi* DTC strategies. Obtained results have been showing in **Figure 7**. It is to be noted that the basic DTC strategy and the DTC-SVM with an imposed commutation frequency ($F_c = 6.25kHz$) strategy offer lower commutation frequencies than the modified *Takahashi* DTC strategy.

3. Concept of PI DTC-SVM based position control

3.1 Computing of flux reference coordinates

The slip angular reference speed ω_r^*, which is the output of the PI controller, will be used to calculate the argument of the stator flux reference. In the reference frame $(\alpha, \ \beta)$, coordinates of the reference stator flux $\phi_{\alpha s}^*$ and $\phi_{\beta s}^*$ are calculated from the polar coordinates according to the following expressions:

$$\begin{cases} \phi_{\alpha s}^* = |\Phi_s^*| \cos\theta_s^* \\ \phi_{\beta s}^* = |\Phi_s^*| \sin\theta_s^* \end{cases} \tag{5}$$

3.2 Computing of voltage reference coordinates

The coordinates of references of voltage vectors $v_{\alpha s}^*$ and $v_{\beta s}^*$ in $(\alpha, \ \beta)$ frame are determined by the following equations:

$$\begin{cases} V_{\alpha s}^* = \dfrac{\phi_{\alpha s}^* - \phi_{\alpha s}}{T_e} + R_s i_{\alpha s} \\ V_{\beta s}^* = \dfrac{\phi_{\beta s}^* - \phi_{\beta s}}{T_e} + R_s i_{\beta s} \end{cases} \tag{6}$$

Finally, they are introduced to the SVM block, which uses them to control the inverter switches (S_a, S_b, S_c).

3.3 Position control loop

The objective is the design of a suitable controller as described by **Figure 8**.

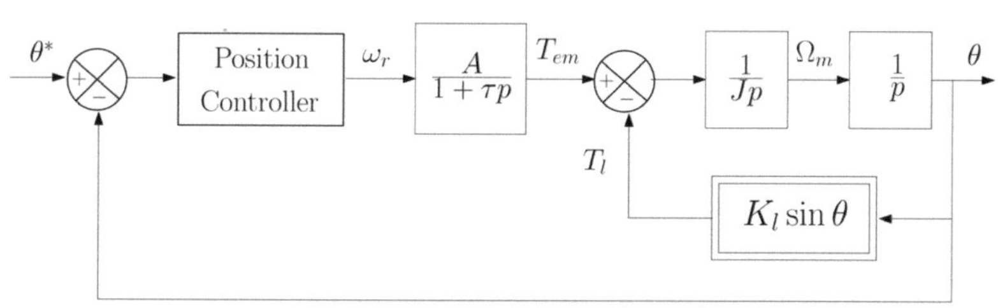

Figure 8.
Position control loop.

Then, we have:

$$\begin{cases} \dfrac{d\theta}{dt} = \Omega_m \\[2ex] \dfrac{d^2\theta}{dt^2} = \dfrac{1}{J}T_{em} - \dfrac{K_l}{J}\sin\theta \\[2ex] \dfrac{d^3\theta}{dt^3} = \dfrac{1}{J}\dfrac{dT_{em}}{dt} - \dfrac{K_l}{J}\dfrac{d\theta}{dt}\cos\theta \end{cases} \tag{7}$$

This yields:

$$\begin{aligned} \frac{d^3\theta}{dt^3} &= \frac{1}{J}\left(-\frac{1}{\tau}T_{em} + \frac{A}{\tau}\omega_r\right) - \frac{K_l}{J}\frac{d\theta}{dt}\cos\theta \\ &= -\frac{1}{\tau}\left(\frac{d^2\theta}{dt^2} + \frac{K_l}{J}\sin\theta\right) + \frac{A}{J\tau}\omega_r - \frac{K_l}{J}\frac{d\theta}{dt}\cos\theta \end{aligned} \tag{8}$$

Thus, we can write:

$$\frac{d^3\theta}{dt^3} + \frac{1}{\tau}\frac{d^2\theta}{dt^2} + \frac{K_l}{J}\frac{d\theta}{dt} + \frac{K_l}{J\tau}\theta + \varphi\left(\theta, \frac{d\theta}{dt}\right) = \frac{A}{J\tau}\omega_r \tag{9}$$

where:

$$\varphi\left(\theta, \frac{d\theta}{dt}\right) = \frac{K_l}{J}\frac{d\theta}{dt}(\cos\theta - 1) + \frac{K_l}{J\tau}(\sin\theta - \theta) = 0\left(\theta, \frac{d\theta}{dt}\right)^3 \tag{10}$$

For small values of θ, $\varphi\left(\theta, \frac{d\theta}{dt}\right)$ can be neglected, and then the mechanical part of the machine can be represented by a third order linear system described by the following transfer function:

$$\frac{\theta}{\omega_r} = \frac{\frac{A}{J\tau}}{\left(p^2 + \frac{K}{J}\right)\left(p + \frac{1}{\tau}\right)} \tag{11}$$

It is to be noted that the application of the following nonlinear feedback represents a nonlinear compensator:

$$\omega_{r'} = \omega_r - \frac{J\tau}{K_l}\varphi\left(\theta, \frac{d\theta}{dt}\right) \tag{12}$$

This loop realizes a feedback linearization. The transfer function between θ and ω_r' is expressed as:

$$\frac{\theta}{\omega_r'} = \frac{\frac{A}{J\tau}}{\left(p^2 + \frac{K}{J}\right)\left(p + \frac{1}{\tau}\right)} \tag{13}$$

which is an exact transfer function without any approximation.

Observing this transfer function, it is clear that it contains two imaginary poles. This leads to a certain difficulty to control the system with a PID controller $C(p)$:

$$C(p) = K_c\left(1 + \frac{1}{T_i p} + T_d p\right) \tag{14}$$

In fact, the system does not present any stability margin. Moreover, to have an adequate dynamical behavior, the derivative time constant (T_d) should be larger that the integral time constant (T_i), which is strongly not recommended.

3.4 DTC-SVM based position control scheme

The implementation scheme of a DTC-SVM based position regulation of an induction motor is shown in **Figure 9**. The idea is based on the decoupling between the amplitude and the argument of the stator flux reference vector.

The amplitude of this vector will be imposed equal to the nominal value of the stator flux, but the argument will be calculated according to the desired performances. In fact, the error between the reference position θ^* and the measured one θ is applied to the position regulator whose output provides the slip angular reference speed ω_r^*, which will be used to calculate the argument of the stator flux reference. Coordinates of the reference stator flux in the reference frame $(\alpha, \ \beta)$ are computed from its polar coordinates according to Eqs. (5). The coordinates of the reference voltage vector $v_{\alpha s}^*$ and $v_{\beta s}^*$ are determined using Eqs. (6).

Finally, the SVM block, which uses these later to generate the convenient stator voltages inverter in each modulation period, ensuring working with a constant commutation frequency.

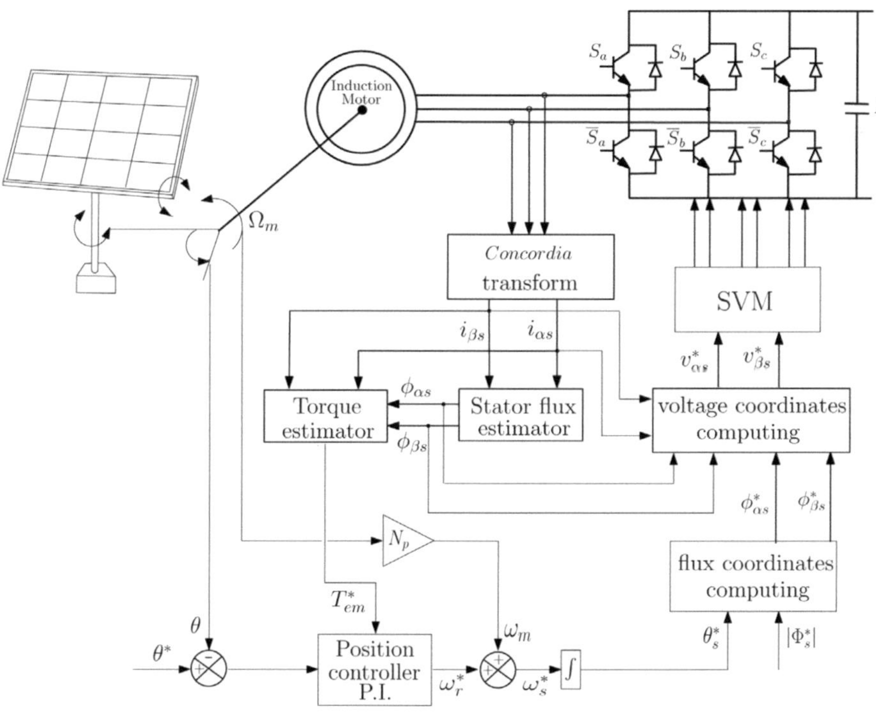

Figure 9.
Induction motor position regulation based on the DTC-SVM strategy.

3.5 Concept of sliding mode DTC-SVM based position control

Sliding mode (SM) controllers perform well in non nonlinear systems than PI controllers [13, 14]. Indeed, the sliding mode control is a type of variable structure systems characterized by the high simplicity and the robustness against insensitivity to parameter variations and external disturbances [14–16]. Considering a nonlinear system described by the following state equation:

$$\dot{X} = f(X) + g(X)U \tag{15}$$

A choice of the sliding surface $S(X)$ can be given by:

$$S(X) = h(X) - h(X^*) \tag{16}$$

with X^* is a reference trajectory.

In order to decide a system trajectory, the equivalent control U_{eq} represents the required control to reach and to remain on the sliding surface. The corrected term ΔU is required to guarantee the remaining on the surface $S(X) = 0$.

Thus, one can choose for the controller the following expression:

$$U = U_{eq} + \Delta U \tag{17}$$

The equivalent control can be designed as follows: when the system remains on the sliding surface, we have $S(X) = 0$, then $\dot{S}(X) = 0$. Since:

$$\dot{S}(X) = h_1(X)[f(X) + g(X)U] - h_1(X^*)\dot{X}^* = \mathcal{F}(X, X^*) + \mathcal{G}(X)U \tag{18}$$

where $h_1(X) = \frac{dh}{dX}$.

This yields the following expression of the equivalent control:

$$U_{eq} = [h_1(X)g(X)]^{-1}\left[h_1(X^*)\dot{X}^* - h_1(X)f(X)\right] = -[\mathcal{G}(X)]^{-1}\mathcal{F}(X, X^*) \tag{19}$$

under the regularity of matrix $\mathcal{G}(X) = [h_1(X)g(X)]$.

The term ΔU can be expressed as:

$$\Delta U = -U_0 \operatorname{sign}\left[\mathcal{G}^T(X)S(X)\right] \tag{20}$$

In fact, if we consider the Lyapunov function:

$$V(X) = S^T S > 0 \tag{21}$$

Its differential with respect to time is expressed as:

$$\dot{V} = S^T\dot{S} = S^T\mathcal{G}(X)\Delta U = -U_0 S^T\mathcal{G}(X)\operatorname{sign}\left[\mathcal{G}^T(X)S(X)\right] = -U_0\left\|\mathcal{G}^T(X)S(X)\right\|_1 \leq 0 \tag{22}$$

This yields that the closed loop system is stable.

3.5.1 Position sliding mode controller

The sliding surface is expressed as:

$$S_\theta = \left(\frac{d}{dt} + \lambda_1\right)^2 \varepsilon_\theta = 0 \tag{23}$$

that is to say:

$$S_\theta = \frac{d^2\varepsilon_\theta}{dt^2} + 2\lambda_1\frac{d\varepsilon_\theta}{dt} + \lambda_1^2\varepsilon_\theta = 0 \tag{24}$$

with: $\varepsilon_\theta = \theta - \theta^*$. This choice takes into account that the error decreases exponentially after reaching the sliding surface. In fact, if $S_\theta = 0$, for $t \geq t_0$, we have: $\varepsilon_\theta(t) = \{\varepsilon_\theta(t_0) + [\dot{\varepsilon}_\theta(t_0) + \lambda_1\varepsilon_\theta(t_0)](t - t_0)\}e^{-\lambda_1(t-t_0)}$.

In this case, function $h(X)$ is expressed as:

$$h(X) = \frac{1}{J}T_{em} + 2\lambda_1\Omega_m + \lambda_1^2\theta - \frac{K_l}{J}\sin\theta \tag{25}$$

To remain the state of the system on the sliding surface $S_\theta = 0$, we have: $\dot{S}_\theta = 0$. This leads to:

$$\dot{S}_\theta = \frac{d^3\varepsilon_\theta}{dt^3} + 2\lambda_1\frac{d^2\varepsilon_\theta}{dt^2} + \lambda_1^2\frac{d\varepsilon_\theta}{dt} = 0 \tag{26}$$

That is to say:

$$\dot{S}_\theta = \frac{1}{J}\left(-\frac{1}{\tau}T_{em} + \frac{A}{\tau}U\right) + 2\lambda_1\left(\frac{1}{J}T_{em} - \frac{K_l}{J}\sin\theta\right) + \left(\lambda_1^2 - \frac{K_l}{J}\cos\theta\right)\Omega_m - h_1(X^*)\dot{X}^* = 0 \tag{27}$$

where:

$$h_1(X^*)\dot{X}^* = \left(\ddot{\Omega}_m^* + 2\lambda_1\dot{\Omega}_m^* + \lambda_1^2\dot{\theta}^*\right) \tag{28}$$

Then, it is easy to express the so-called equivalent control which corresponds to the required control remaining the system on the sliding surface:

$$U_{eq,\theta} = \frac{1}{A}T_{em} - 2\frac{\lambda_1\tau}{A}(T_{em} - K_l\sin\theta) - \frac{J\tau}{A}\lambda_1^2\Omega_m + \frac{\tau K_l}{A}\Omega_m\cos\theta + \frac{J\tau}{A}h_1(X^*)\dot{X}^* \tag{29}$$

Then, the slip angular reference speed ω_r can be expressed by:

$$\omega_r = U_{eq,\theta} - U_{0,\theta}\text{sign}\ (S_\theta) \tag{30}$$

The new structure of this control approach is given by the block diagram of **Figure 10**.

3.6 SM controllers with adaptive parameters estimation

If system (14) depends on an unknown parameter vector $\gamma = [\gamma_1\gamma_2 \ldots]^T$, the expression of the control depends on γ, that is to say: $U_{eq} = U_{eq}(\gamma)$ and the applied control law becomes:

$$\overline{U} = \overline{U}_{eq} + \Delta U \tag{31}$$

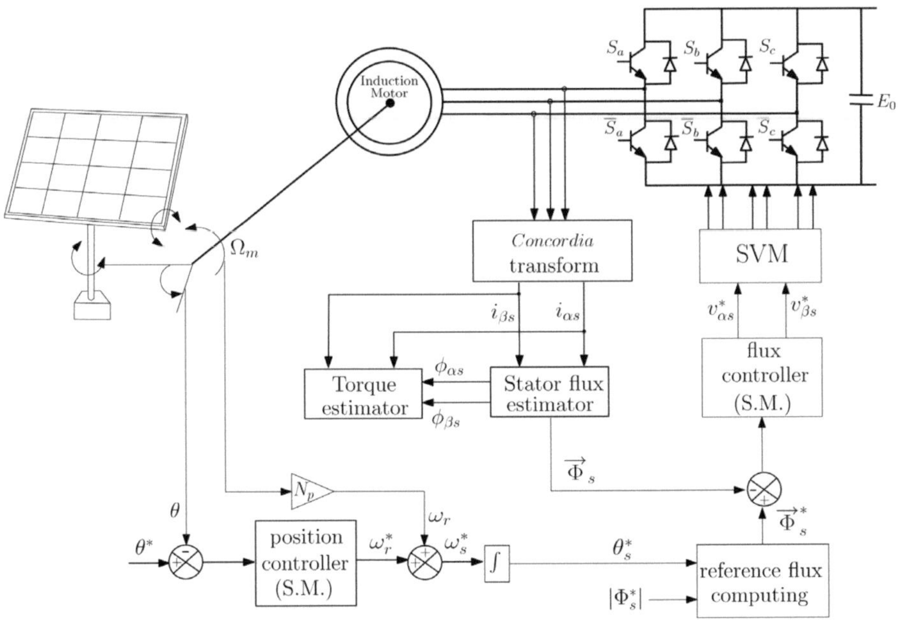

Figure 10.
Induction motor position regulation based on the DTC-SVM with sliding mode controllers.

where $\overline{U}_{eq} = U_{eq}(\overline{\gamma})$. $\overline{\gamma}$ is the estimated vector of γ.

Referring to Eq. (17), the differential of S is expressed as:

$$
\begin{aligned}
\dot{S} &= \mathcal{F}(X,X^*) + \mathcal{G}(X)\overline{U} = \mathcal{F}(X,X^*) + \mathcal{G}(X)\left(\overline{U}_{eq} + \Delta U\right)\\
&= \underbrace{\mathcal{F}(X,X^*) + \mathcal{G}(X)U_{eq}}_{=0} + \mathcal{G}(X)\Delta U + \mathcal{G}(X)\left(\overline{U}_{eq} - U_{eq}\right)\\
&= \mathcal{G}(X)\Delta U + \mathcal{G}(X)\left(\overline{U}_{eq} - U_{eq}\right)\\
&= \mathcal{G}(X)\Delta U + \mathcal{G}(X)\left(\overline{U}_{eq} - U_{eq}\right) + \underbrace{\underbrace{\left[\mathcal{G}(X) - \overline{\mathcal{G}}(X)\right]}_{=o(\Delta\gamma)}\underbrace{\left(\overline{U}_{eq} - U_{eq}\right)}_{=o(\Delta\gamma)}}_{=o(\Delta\gamma)^2}\\
&= \mathcal{G}(X)\Delta U + \overline{\mathcal{G}}(X)\left(\Sigma_i \frac{\partial U_{eq}}{\partial\gamma_i}(\overline{\gamma})\Delta\gamma_i\right) + o(\Delta\gamma)^2\\
&= \overline{\mathcal{G}}(X)\Delta U + \overline{\mathcal{G}}(X)\left(\Sigma_i \frac{\partial U_{eq}}{\partial\gamma_i}(\overline{\gamma})\Delta\gamma_i\right) + \left[\mathcal{G}(X) - \overline{\mathcal{G}}(X)\right]\Delta U + o(\Delta\gamma)^2\\
&= \overline{\mathcal{G}}(X)\Delta U + \overline{\mathcal{G}}(X)\left(\Sigma_i \frac{\partial U_{eq}}{\partial\gamma_i}(\overline{\gamma})\Delta\gamma_i\right) + \left[\mathcal{G}(X) - \overline{\mathcal{G}}(X)\right]\Delta U + o(\Delta\gamma)^2\\
&= \overline{\mathcal{G}}(X)\Delta U + \left(\sum_i \left[\overline{\mathcal{G}}(X)\frac{\partial U_{eq}}{\partial\gamma_i}(\overline{\gamma}) + \frac{\partial\mathcal{G}(X)}{\partial\gamma_i}\Delta U\right]\Delta\gamma_i\right) + o(\Delta\gamma)^2
\end{aligned}
\tag{32}
$$

with $\Delta\gamma = \overline{\gamma} - \gamma$ and $\overline{\mathcal{G}}(\mathrm{X})$ is the expression of $\mathcal{G}(\mathrm{X})$ for $\gamma = \overline{\gamma}$.

- Theorem

Control laws (17), (19) and (20) stabilize system (15) with the following adaptive laws:

$$\dot{\bar{\gamma}}_i = -\eta_i S^T\left(\overline{\mathcal{G}}(X)\frac{\partial U_{eq}}{\partial \gamma_i}(\bar{\gamma}) + \frac{\partial \mathcal{G}(X)}{\partial \gamma_i}\Delta U\right) \tag{33}$$

- Proof

Let us consider the following Lyapunov function:

$$V = \frac{1}{2}S^T S + \frac{1}{2}\sum_i \frac{1}{\eta_i}\Delta\gamma_i^2 \tag{34}$$

In the following, it assumed that vector γ is constant or it has slow variations with respect to time, in such away that we can neglect its differential with.

respect to time: $\dot{\gamma} \simeq 0$. Then, we can write: $\Delta\dot{\gamma} \simeq \dot{\bar{\gamma}}$.

The differential with respect to time of function V is expressed as:

$$\begin{aligned}
\dot{V} &= s^T\dot{S} + \Sigma_i\frac{1}{\eta_i}\Delta\gamma_i\Delta\dot{\gamma}_i \\
&= S^T\left[\overline{\mathcal{G}}(X)\Delta U + \left(\Sigma_i\left[\overline{\mathcal{G}}(X)\frac{\partial U_{eq}}{\partial \gamma_i}(\bar{\gamma}) + \frac{\partial g(X)}{\partial \gamma_i}\Delta U\right]\Delta\gamma_i\right) + o(\Delta\gamma)^2\right] \\
&\quad + \Sigma_i\frac{1}{\eta_i}\Delta\gamma_i\Delta\dot{\gamma}_i + o(\Delta\gamma)^2 \\
&= S^T\overline{\mathcal{G}}(X)\Delta U + \Sigma_i\underbrace{\left[S^T\left(\overline{\mathcal{G}}(X)\frac{\partial U_{eq}}{\partial \gamma_i}(\bar{\gamma}) + \frac{\partial \mathcal{G}(X)}{\partial \gamma_i}\Delta U\right) + \frac{1}{\eta_i}\Delta\dot{\gamma}_i\right]}_{=0}\Delta\gamma_i + o(\Delta\gamma)^2 \\
&= S^T\overline{\mathcal{G}}(X)\Delta U + o(\Delta\gamma)^2 = -S^T\overline{\mathcal{G}}(X)\,\text{sign}\left[\overline{\mathcal{G}}^T(X)S(X)\right] + o(\Delta\gamma)^2 \\
&= -\left\|\overline{\mathcal{G}}^t(X)S(X)\right\|_1 + o(\Delta\gamma)^2 \le 0
\end{aligned} \tag{35}$$

where $\|\cdot\|_1$ is the norm "1" of a vector which corresponds to the sum of absolute values of its components.

3.6.1 Position adaptive SM controller with variations on the mutual inductance and the rotor resistance

The sensitivity of the DTC-SVM to (i) variations on the magnetic permeability of the stator and rotor cores, and (ii) variations on the rotor resistance, which can vary with time and operating conditions, can be removed by an online estimation of the mutual inductance and the rotor resistance. The adaptive SM of the speed can be derived based on the mutual inductance and rotor resistance estimations using the *Lyapunov* theorem [17].

It is easy to show that:

$$U_\theta = \overline{U}_{eq,\theta} - U_{0,\theta}\ \text{sign}\ (S_\Omega) \tag{36}$$

where:

$$\overline{U}_{eq,\theta} = -\left[\overline{\mathcal{G}}(X)\right]^{-1}\overline{\mathcal{F}}(X,X^*) \tag{37}$$

Then:

$$\begin{aligned}\dot{S}_\theta &= \mathcal{F}(X,X^*) + \mathcal{G}(X)\overline{U}_\theta = \mathcal{F}(X,X^*) + \mathcal{G}(X)\left(\overline{U}_{eq,\theta} + \Delta U_\theta\right) \\ &= \mathcal{F}(X,X^*) + \mathcal{G}(X)U_{eq,\theta} + \mathcal{G}(X)\Delta U_\theta + \mathcal{G}(X)\left(\overline{U}_{eq,\theta} - U_{eq,\theta}\right) \\ &= \mathcal{G}(X)\Delta U_\theta + \mathcal{G}(X)\left(\overline{U}_{eq,\theta} - U_{eq,\theta}\right)\end{aligned} \tag{38}$$

- Corollary

The following slip angular reference speed control law stabilizes the speed loop:

$$\omega_r = \overline{U}_{eq,\theta} - U_{0,\theta}\ \text{sign}\ \ (S_\Omega) \tag{39}$$

where $\overline{U}_{eq,\theta} = U_{eq,\theta}\left(\overline{M},\overline{R}_r\right), \overline{M}$ and $\overline{R}_r$ are estimator values of the mutual inductance and the rotor resistance given by the following updating laws:

$$\begin{aligned}\dot{\overline{M}} &= -\eta_{\theta_1}\overline{\mathcal{G}}S_\theta\left(\frac{\partial U_{eq,\theta}}{\partial M}\right) \\ \dot{\overline{R}}_r &= -\eta_{\theta_2}\overline{\mathcal{G}}S_\theta\left(\frac{\partial U_{eq,\theta}}{\partial R}\right)\end{aligned} \tag{40}$$

with: η_{θ_1} and η_{θ_2} positive scalars, $\mathcal{G} = \mathcal{G}(M,R_r)$ and $\overline{\mathcal{G}} = \overline{\mathcal{G}}\left(\overline{M},\overline{R}_r\right)$ defined in Eq. (17).

- Proof

Considering the following function:

$$V_\theta = \frac{1}{2}S_\theta^2 + \frac{1}{2\eta_{\theta_1}}\Delta M^2 + \frac{1}{2\eta_{\theta_2}}\Delta R_r^2 \tag{41}$$

with $\Delta M = \overline{M} - M$ and $\Delta R_r = \overline{R}_r - R_r$.
The time derivative of the *Lyapunov* function can be expressed as:

$$\dot{V}_\theta = S_\theta\dot{S}_\theta + \frac{1}{\eta_{\theta_1}}\Delta M\Delta\dot{M} + \frac{1}{\eta_{\theta_2}}\Delta R_r\Delta\dot{R}_r \tag{42}$$

However:

$$\overline{U}_{eq,\theta} - U_{eq,\theta} = \left(\overline{M} - M\right)\left(\frac{\partial U_{eq,\theta}}{\partial M}\right) + \left(\overline{R}_r - R_r\right)\left(\frac{\partial U_{eq,\theta}}{\partial R_r}\right) + o(\Delta M,\ \ \Delta R_r)^2 \tag{43}$$

Moreover:

$$\mathcal{G}\ \left(\overline{U}_{eq,\theta} - U_{eq,\theta}\right) = \overline{\mathcal{G}}\left(\overline{U}_{eq,\theta} - U_{eq,\theta}\right) + o(\Delta M,\ \ \Delta R_r)^2 \tag{44}$$

DOI: http://dx.doi.org/10.5772/intechopen.94436

Thereby, Eq. (42) gives:

$$
\begin{aligned}
\dot{V}_\theta &= -U_{0,\theta} \mathcal{G}(X)|S_\Omega| + \Delta M \underbrace{\left[\overline{\mathcal{G}} \left(\frac{\partial U_{eq,\theta}}{\partial M} \right) + \frac{1}{\eta_{\theta_1}} \Delta \dot{M} \right]}_{=0} \\
&+ \Delta R_r \underbrace{\left[\overline{\mathcal{G}} \left(\frac{\partial U_{eq,\theta}}{\partial R_r} \right) + \frac{1}{\eta_{\theta_2}} \Delta \dot{R}_r \right]}_{=0} + o(\Delta M, \Delta R_r)^2 \\
&= -U_{0,\theta} \mathcal{G}(X)|S_\Omega| + o(\Delta M,\ \Delta R_r)^2 \leq 0
\end{aligned}
\tag{45}
$$

Since $\mathcal{G}(X) > 0$, $\dot{V}_\theta$ is negative. Then, the system is stable.

3.7 Simulation results investigated SM DTC-SVM approach based position control

Simulation works have been carried out in order to investigate performances of the position control of the induction motor drive under the above-presented DTC-SVM strategies, using PID, PID with a nonlinear compensator and SM controllers. For the sake of comparison, both strategies have been considered in the same induction motor drive using the same implementation conditions, such that:

- a reference stator flux Φ_s^* equal to 1 Wb,
- The modulation period has been fixed to $T_{mod} = 150\mu s$ in all DTC-SVM approaches under study,
- Constants involved in the position SM controller are: $U_{0,\theta} = 50$ and $\lambda_\theta = 100$. The constants involved in the flux SM controller are: $U_{0,\phi}, = 150$ and $\lambda_\phi = 2$.

The desired trajectory is defined by smooth variations of the position θ, the speed Ω_m and the torque C_{em}, leading to:

- variations of θ form $-60°$ (morning panel position) to $30°$ from 0 s to 1 s,
- constant value of θ equal to $30°$ from 1 s to 1.5 s.
- variations of θ form $30°$ to $60°$ (afternoon panel position) from 1.5 s to 2.5 s,
- constant value of θ equal to $60°$ from 2.5 s to 4 s.

The analysis of simulation results leads to the following items:

- **Figures 11** and **12** present evolutions of the position θ, the speed Ω_m, the torque C_{em}, the flux $|\Phi_s|$ and the current i_{as}, using PID controllers (figures indexed by 1), PID controllers with a nonlinear compensator (figures indexed by 2), and SM controllers (figures indexed by 3). It is well obvious that a good tracking has been realized by these control approaches. It is also obvious, that there is no significant difference between results yielded by PID controllers and PID controllers with a nonlinear compensator. This justifies that the nonlinear term $\varphi\left(\theta, \frac{d\theta}{dt}\right)$ can be neglected. Moreover, ripples of the torque, the flux and stator currents are smallest for results given by SM controllers.

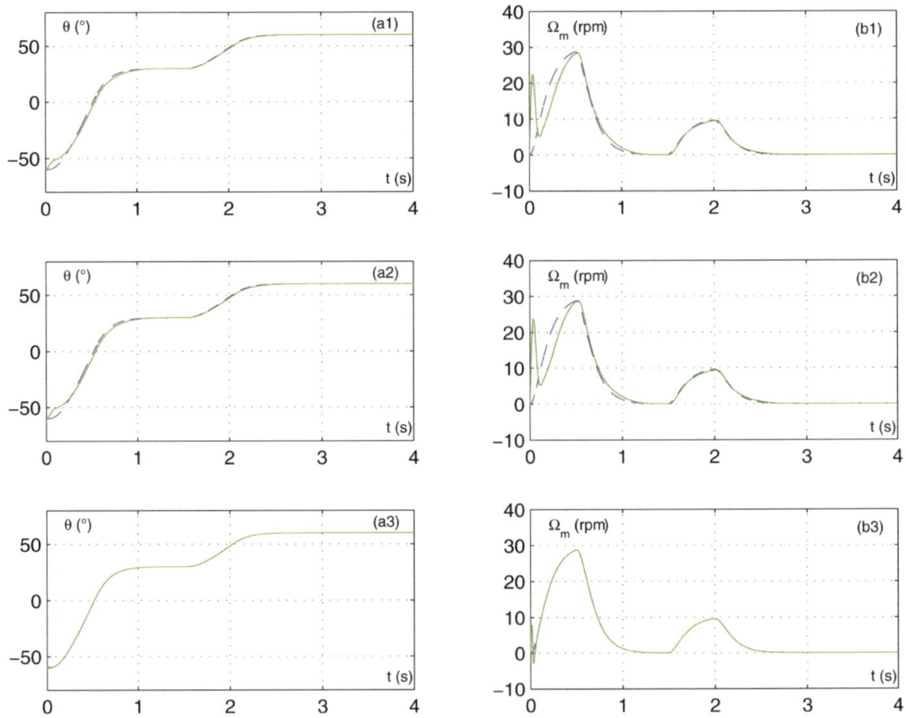

Figure 11.
Induction motor position regulation considering (subscript "1") DTC-SVM approach using PI controller, (subscript "2") DTC-SVM approach using PI controller with a nonlinear compensator and (subscript "3") DTC-SVM approach using sliding mode controllers. Legend: (a) evolution of the position and its reference and (b) the speed of the motor and its reference.

- **Figures 13** and **14** present the same variable evolutions for variations of machine parameters as: +100% variations on the stator resistance R_s, $+100\%$ variations on the rotor resistance R_r and -50% variations on the mutual inductance M. It is clear that results, yielded from PID controllers without and with a nonlinear compensator, present important oscillations. However, SM controllers with parameter's updating give same results as in the case where parameters are known and do not vary.

- Thus, the implementation of the DTC-SVM using sliding mode controllers strategies highlights high dynamical performances obtained with the lowest torque ripple, the lowest flux ripple and the lowest current ripple.

3.8 Performance criteria

Considering the same simulation, we propose to use performance criteria defined in the appendix.

In the following, the steady state operating point is defined by a desired position θ equal to 60° for the time larger than 2.5 s.

- Total Harmonic Distortion (THD)

The first criterion is the average total harmonic distortion (THD) of the stator current which is defined in the Appendix.

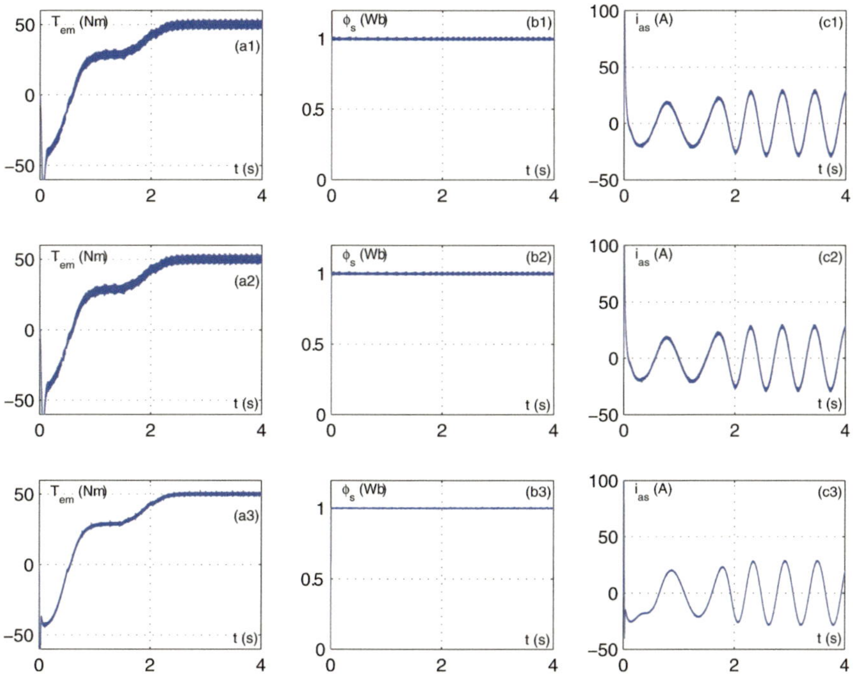

Figure 12.
Induction motor position regulation considering (1) DTC-SVM approach using PI controller, (2) DTC-SVM approach using PI controller with a nonlinear compensator and (3) DTC-SVM approach using sliding mode controllers. Legend: (a) evolution of the electromagnetic torque, (b) the stator flux and (c) the stator current of phase a.

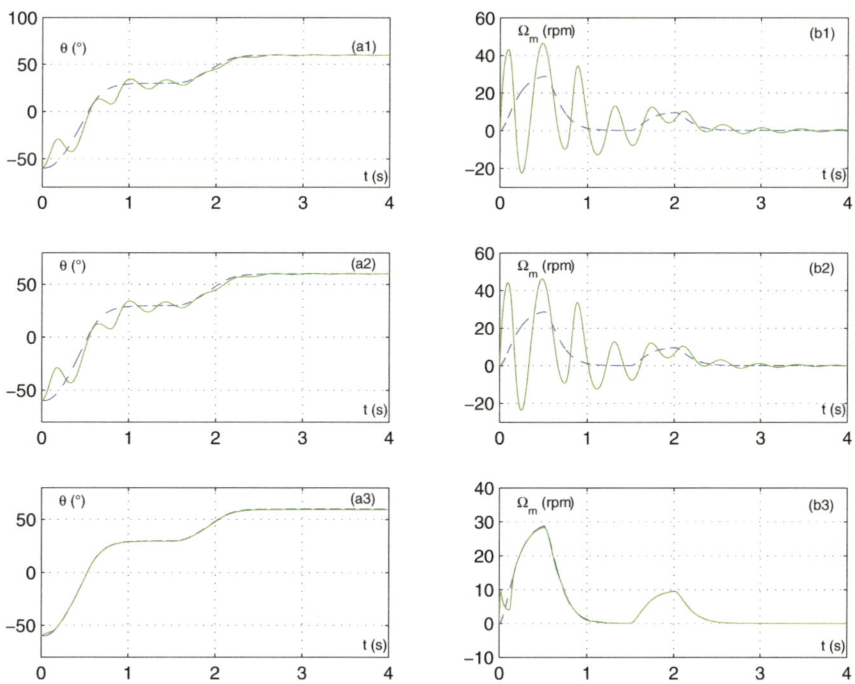

Figure 13.
Induction motor position regulation, considering +100% variations on the stator resistance, (1) DTC-SVM approach using PI controller, (2) DTC-SVM approach using PI controller with a nonlinear compensator and (3) DTC-SVM approach using sliding mode controllers. Legend: (a) evolution of the position and its reference and (b) the speed of the motor and its reference.

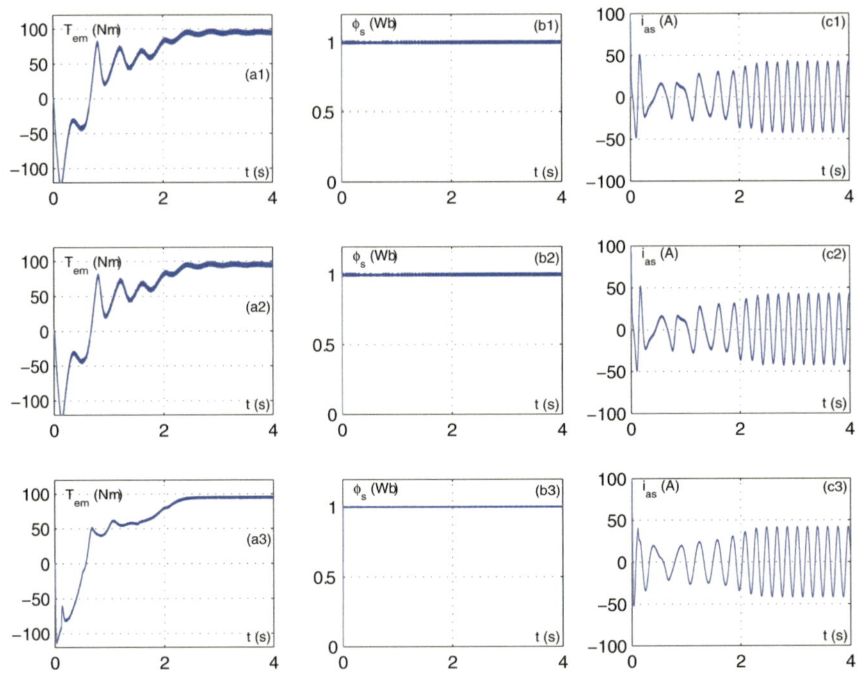

Figure 14.
Induction motor position regulation, considering +100% variations on the stator resistance, considering (1) DTC-SVM approach using PI controller, (2) DTC-SVM approach using PI controller with a nonlinear compensator and (3) DTC-SVM approach using sliding mode controllers. Legend: (a) evolution of the electromagnetic torque, (b) the stator flux and (c) the stator current of phase a.

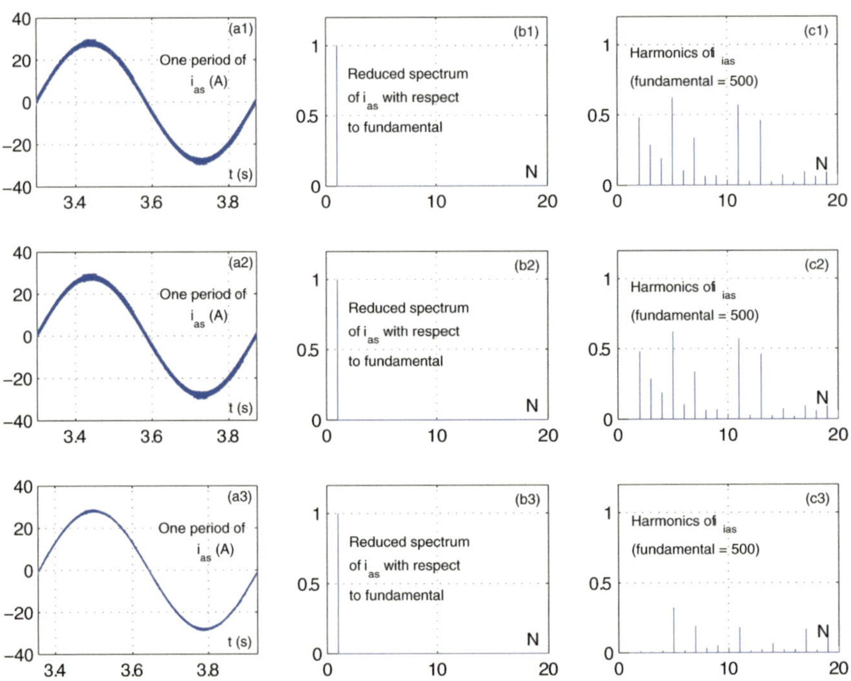

Figure 15.
Spectrum of the current i_{as}. (a) Normalized spectrum, (b) higher harmonics of the spectrum current (c) one period of the current i_{as}. (subscript "1") DTC-SVM approach using PI controller, (subscript "2") DTC-SVM approach using PI controller with a nonlinear compensator and (subscript "3") DTC-SVM approach using sliding mode controllers.

	PI without a NL compensator	PI with a NL Compensator	Sliding Mode Controllers
THD (%)	3.07	3.07	1.23

Table 4.
Total harmonic distortion of the stator current i_{as}.

In this context, the frequency spectrum of the stator current i_{as} has been analyzed by the observation of amplitudes of all its harmonics frequencies. **Figure 15** shows the evolution of one period of i_{as} between 3 s and 4 s, its spectrum (only 20 harmonics has been presented). It is obvious that SM controllers give less ripples of the stator current.

The total harmonic distorsion criterion of the stator current i_{as} is given by **Table 4** which shows that SM controllers give the lowest criterion.

- Ratio of torque and flux ripples.

The second comparison criterion translates the torque and the flux ripples around their steady state values $|\Phi_s| = 1$ and $T_{em,mean} = K_l \sin \frac{\pi}{6}$.

Figure 16 presents the evolution of the torque T_{em} and the flux $|\Phi_s|$ from 3 s to 4 s. Computations of flux ripple criteria are given by **Table 5**, and computations of torque ripple criteria are given by **Table 6**. These tables confirm that the PID controllers without a nonlinear compensator and PID controllers with a nonlinear compensator give same results. However, SM controllers give less ripples of the flux and the torque.

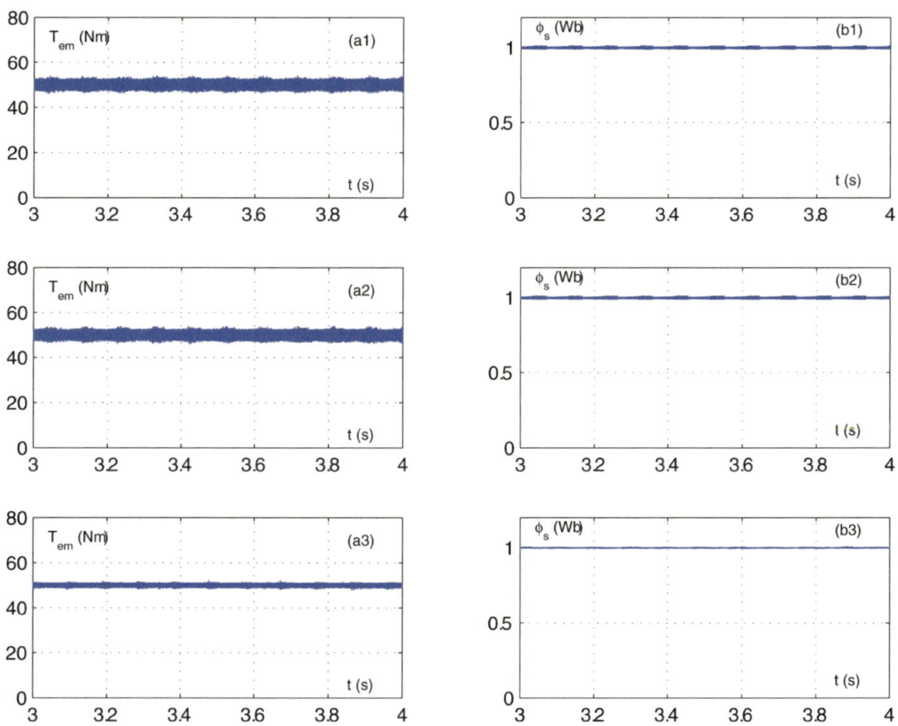

Figure 16.
Zoomed shapes of (a) electromagnetic torque and (b) stator flux. In the case of (subscript "1") DTC-SVM approach using PI controller, (subscript "2") DTC-SVM approach using PI controller with a nonlinear compensator and (subscript "3") DTC-SVM approach using sliding mode controllers.

	PI without a NL compensator	PI with a NL Compensator	Sliding Mode Controllers
$\Phi_{RIP,1}$ (%)	0.38	0.38	0.12
$\Phi_{RIP,2}$ (%)	0.44	0.44	0.15
$\Phi_{RIP,\infty}$ (%)	1.26	1.33	0.65

Table 5.
Flux ripple criteria.

	PI without a NL compensator	PI with a NL Compensator	Sliding Mode Controllers
$T_{RIP,1}$ (%)	1.88	1.86	0.92
$T_{RIP,2}$ (%)	2.71	2.66	0.62
$T_{RIP,\infty}$ (%)	8.17	8.34	4.65

Table 6.
Torque ripple criteria.

4. Conclusion

In this chapter, the DTC position control of induction motor controlling photovoltaic panel has been considered. This panel is commonly exposed to the sun in fixed positions corresponding to the maximum sunshine recorded during a day. Firstly, the DTC-SVM approach using hysteresis controllers has been compared to the basic DTC strategy and DTC strategy with a look-up table including only active voltage vectors. Then, the problem of position regulation of an IM under DTC-SVM approaches has been treated. In fact, a comparison between three DTC-SVM approaches: a DTC-SVM approach using PI controllers, a DTC-SVM approach using PI controllers with a nonlinear compensator, and a DTC-SVM approach using sliding mode controllers, has been proposed. Finally, an adaptation approach of parameter estimators has been implemented in order to eliminate the effects of parameter variations and load disturbances. It has been shown through simulations the sliding mode DTC-SVM approach (i) eliminates the demagnetization effects, and gives lowest ripples on the torque and on the flux, (ii) presents less harmonic distortion on the stator currents, and (iii) it presents good performances with a good robustness with respect to parameter's variations and load disturbances, particularly in the case of adapted estimators of machine parameters.

Author details

Fatma Ben Salem
Control and Energy Management Laboratory (CEMLab), University of Sfax, Sfax Engineering School, BP 1173, 3038 Sfax, Tunisia

*Address all correspondence to: fatma.bensalem@isgis.usf.tn; fatma_bs@yahoo.fr

IntechOpen

© 2020 The Author(s). Licensee IntechOpen. This chapter is distributed under the terms of the Creative Commons Attribution License (http://creativecommons.org/licenses/by/3.0), which permits unrestricted use, distribution, and reproduction in any medium, provided the original work is properly cited.

References

[1] Takahashi., I., and Ohmori, Y., High Performance Direct Torque Control of Induction Machine. IEEE Power Electronics Specialists Conf. (PESC'88), 2:870–876, Kyoto-Japan, April 1988.

[2] Depenbrock, M., 'Direct Self-Control (DSC) of Inverter-Fed Induction Machine'. *IEEE Trans. Industry Applications,* 28(3):581–588, May-juin 1992.

[3] Ben Salem, F., and Feki, M., (2019) 'An improved DTC of Induction Motor for Electric Vehicle Propulsion: An Intent to Provide Comfortable Ride', *from book: Optimization of Green Transportation Problems: Fundamentals and Applications,* Wiley, December 16, 2019.

[4] El Ouanjli, N., Derouich, A., El Ghzizal, A., Motahhir, S., Chebabhi, A., El Mourabit, Y., and Taoussi, M., (2019) 'Modern improvement techniques of direct torque control for induction motor drives - a review', *Protection and Control of Modern Power Systems,* 4(11): 12 pages.

[5] Bouzidi, B., Ben Salem, F., Yangui, A., and Masmoudi,A., (2007) 'Direct Torque Control strategy Based Maximum Sunshine Position Tracking', *Second International Conference and Exhibition on Ecological Vehicles and Renewable Energies* (EVER'07), Monte Carlo-Monaco, March–April 2007.

[6] Bouzidi, B., Ben Salem, F., and Yangui, A., (2008) 'The Classical and Analytic DTC for Photovoltaic Panel Position and Control', *International Journal of Sciences and Techniques of Automatic Control and Computing Engineering* (IJ-STA), special issue CEM, 636–651.

[7] Davari, S. A., Hasankhan, E. and haburi, D. A. (2011) 'A comparative study of DTC-SVM with three-level inverter and an improved predictive torque control using two-level inverter', *2nd Power Electronics, Drive Systems and Technologies Conference,* Tehran, Iran.

[8] Zhifeng, Z., Renyuyan,T., Boadong, B. and Dexin, X. (2010) 'Direct torque control based on space vector modulation with adaptative stator flux observer for induction motors', *IEEE Trans. Magnetics,* Vol. 48, No. 8, pp.3133–3136.

[9] Meroufel, A., Wira, P., Nefsi, M. and Massoum, A. (2012) 'Contle directe du couple de la machine asynchrone bas sur MLI vectorielle discrtise (DSVM-DTC)', *Acta Electrotehnica*, Vol. 1, pp.35–40.

[10] Abu-Rub, H., Stando, D. and Kazmierkowski, M.P. (2013) 'Simple speed sensorless DTC-SVM scheme for induction motor drives', *Bulletin of the Polish Academy of Sciences Technical Sciences,* Vol. 61, No. 2, pp.301–307.

[11] Manuel, A. and Francis, J. (2013)'Simulation of direct torque controlled induction motor drive by using space vector pulse width modulation for torque ripple reduction', *International Journal of Advanced Research in Electrical, Electronics and Instrumentation Engineering*, Vol. 2, No. 9, pp.4471–4478.

[12] Bendaikha, A., Saad, S., Abdou, A., Defdaf, M., Laamari Y., (2019) 'A Study of SVM-DTC and Conventional DTC for Induction Motors Drive Fed by Five-level Inverter', *European Journal of Electrical Engineering,* Vol. 21, No. 1, pp.85–91.

[13] Ben Salem,F., Bahri,I., Maamri,H. and Derbel,N. A Second-Order Sliding Mode Control of Switched Reluctance Motor, J. of Power and Energy Systems, 2020, doi.org/10.1080/ 15325008.2020.1797937, pp. 640–651.

[14] Ben Salem,F. and Derbel N., Direct torque control of induction motors based on discrete space vector modulation using adaptive sliding mode control, Int. J. of Electric Power Components and Systems, 2014, 42, (14), pp. 1598–1610.

[15] Ben Salem,F. and Derbel,N., A Sliding Mode field Oriented Control of an Induction Machine Operating with Variable Parameters, J. of Power and Energy Systems, 2007, 27,(2), pp. 205212.

[16] Ben Salem,F. and Derbel,N., DTC-SVM-Based Sliding Mode Controllers with Load Torque Estimators for Induction Motor Drives, Chapter 14 of the Book: Applications of Sliding Mode Control, Studies in Systems, Decision and Control 79, Springer Science +Business Media Singapore 2017, pp. 269297.

[17] F. Ben Salem and N. Derbel, “Position Control Performance Improvement of DTC-SVM for an Induction Motor: Application to Photovoltaic Panel Position”, *International Journal of Renewable Energy Research (IJRER),* 4(4):879–892, 2014.

Section 2

Direct Torque Control Improvements

Chapter 4

Flux Reversal Machine Design

Yuting Gao and Yang Liu

Abstract

Flux reversal permanent magnet machines (FRPMMs) have a simple reluctance rotor and a stator with armature windings and permanent magnets (PMs). Due to the high torque density and high efficiency of FRPMMs, they have been widely used in many applications such as electric vehicle, wind power generation, etc. However, the general design method of FRPMMs has not been established in books. Therefore, this chapter will focus on introducing an analytical design method, which allows for fast design of FRPMMs. First of all, the analytical sizing equations are deduced based on a magneto motive force (MMF)-permeance model. After that, the effects of some key performances including average torque, pulsating torque, power factor, and PM demagnetization are analyzed. Moreover, the feasible slot-pole combinations are summarized and the corresponding winding type of each combination is recommended in order to maximize the output torque. Besides, the detailed geometric design of stator and rotor are presented. Finally, a case study is presented to help readers better understand the introduced design methodology.

Keywords: design method, flux reversal permanent magnet machine (FRPMM), sizing equation, finite element analysis (FEA)

1. Introduction

The topology of FRPMM is depicted in **Figure 1**. As can be seen, it has a slotted rotor without any windings or PMs, and a stator with armature windings and PMs mounted on each stator teeth. First of all, the structural characteristics of FRPMMs and the corresponding performance advantages need to be explained:

1. FRPMMs are excited by PMs instead of the excitation windings, which are different with asynchronous motors and brushed DC motors. So, for FRPMMs, the rotor will not have copper losses, and the efficiency is relatively higher [1, 2].

2. The rotor of FRPMMs has no windings or permanent magnets, thus is suitable for high-speed operation and high-temperature operating conditions [3]. Moreover, the no excitation winding will keep away from the problems of friction noise and electric spark. So, FRPMMs are more reliable and require less maintenance [4, 5]. In addition, the rotor of FRPMMs is light in weight and has a small rotational inertia [6]; hence, the acceleration and deceleration response is faster.

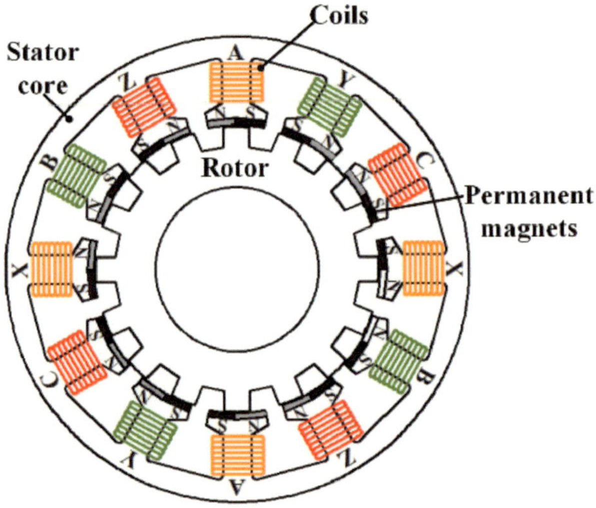

Figure 1.
Cross section of a FRPMM.

3. The stator windings of FRPMMs are mostly concentrated windings, which are easy to manufacture. Moreover, the electromagnetic isolation of the concentrated windings is better than regular distributed windings, which means that if one winding has faults, the fault is not likely to spread to other windings, and thus the fault tolerance is good [7, 8]. In addition, the concentrated winding has a smaller winding factor, inductance, and a shorter electrical time constant than the distributed windings [9], and thus the dynamic response of concentrated winding is faster.

4. Compared to other stator-PM machines, that is, flux switching PM machines and doubly salient PM machines, FRPMMs have a simpler structure. The PMs of the flux switching PM machines and doubly salient PM machines are inserted into the stator core, which is not convenient for installation. In the flux switching PM machine, putting permanent magnets in the middle of the teeth will reduce the slot area and affect the output torque. In the doubly salient PM machine, placing PMs in the yoke will increase the volume of the motor and reduce the torque density. In the FRPMMs, the PMs are pasted on the inner surface of the stator teeth, thus eliminating the above problems [10].

Finally, the structural characteristics and performance advantages of the flux-reverse motor can be summarized in **Table 1**.

It can be seen that the FRPMMs have many performance advantages, and these advantages can be utilized in different applications. First of all, the high efficiency, the large torque density, the rapid acceleration, and deceleration response make FRPMMs suitable for various high-speed rotation areas, such as electric vehicles [11–13], electric spindle [14], fans [15, 16], etc. Secondly, the number of rotor pole pairs is usually high, which is also suitable for low-speed areas, meanwhile its torque density is high at the low speeds, making FRPMMs suitable for various low-speed direct-drive occasions [17], for example wind power [18–20], direct drive servo system [21], wave power generation [22], etc. In addition, linear FRPMM has no PMs and copper windings in the secondary, which saves cost and is also very suitable for long rail transit linear motion applications [23, 24].

No.	Structural characteristic	Advantages
1	Use rare-earth PMs	1. No excitation loss, high motor efficiency; 2. Rare-earth with high-magnetic energy product increases torque density
2	No windings or PMs in the rotor	1. Simple rotor structure, suitable for high-speed operation and high temperature conditions; 2. Avoids mechanical friction and electric sparks caused by commutators and brushes, thus improving the reliability; 3. Small rotational inertia, thus fast acceleration and deceleration response
3	Often use concentrated windings	1. Good fault tolerance and high reliability; 2. Easy processing and manufacturing; 3. Small inductance and electrical time constant
4	PMs attached to the stator teeth surface	1. Easy to install 2. No reduction in the slot area or increase in the motor volume

Table 1.
Structural characteristics and corresponding advantages of FRPMM.

In most existing literatures, the design of FRPMMs is mainly based on the classical design method [25] with low accuracy or time-consuming finite element algorithm (FEA) [26]. Therefore, in this chapter, the specialized sizing equations for FRPMMs will be deduced and the analytical design method will be introduced, which can be directly employed in the initial design of FRPMMs and allows for fast calculations of machine dimensions.

This chapter is organized as follows. First, the structure and operation principles are introduced in Section 2. Then in Section 3, the magnetic circuit model is built and the sizing equations are analytically derived. After that, in Section 4, the influences of several key parameters (slot-pole combination, airgap radius, electric loading, and equivalent magnetic loading) in the sizing equation on the torque density are analyzed. Also, the effects of the airgap structural parameters on the pulsating torque, power factor, and PM demagnetization performances are investigated. Moreover, in Section 5, the geometric design of stator and rotor are introduced. And in Section 6, the design procedure is illustrated. Besides, to make the analytical design method more readable, a case study is presented and a FRPMM prototype is tested. Finally, conclusions are drawn in Section 7.

2. Operation principle of FRPMM

To clearly exhibit the operating principle, a three-phase FRPMM with two pole windings, six stator slots, and eight rotor teeth is cited as an example. The flux distributions at different rotor positions are illustrated in **Figure 2**. The magnetic flux field is excited only by the PMs, and the difference of each rotor movement is 11.25 mech. degrees (i.e., 1/4 rotor slot pitch). Taking flux linkage of phase A winding as an example, when the rotor position is 0 degree, the flux linkage is 0; when the rotor position is 11.25 mech. degree (90 elec. degree), the flux linkage reaches the positive maximum value; when the rotor position is 22.5 mech. degree (180 elec. degree), the flux linkage is 0; when the rotor position is 33.75 mech. degree (270 elec. degree), the flux linkage reaches the negative maximum value. Therefore, in the duration of one rotor slot pitch (360 elec. degrees), the winding flux linkage reverses the polarity, thus it is called “flux reversal machine.” Then,

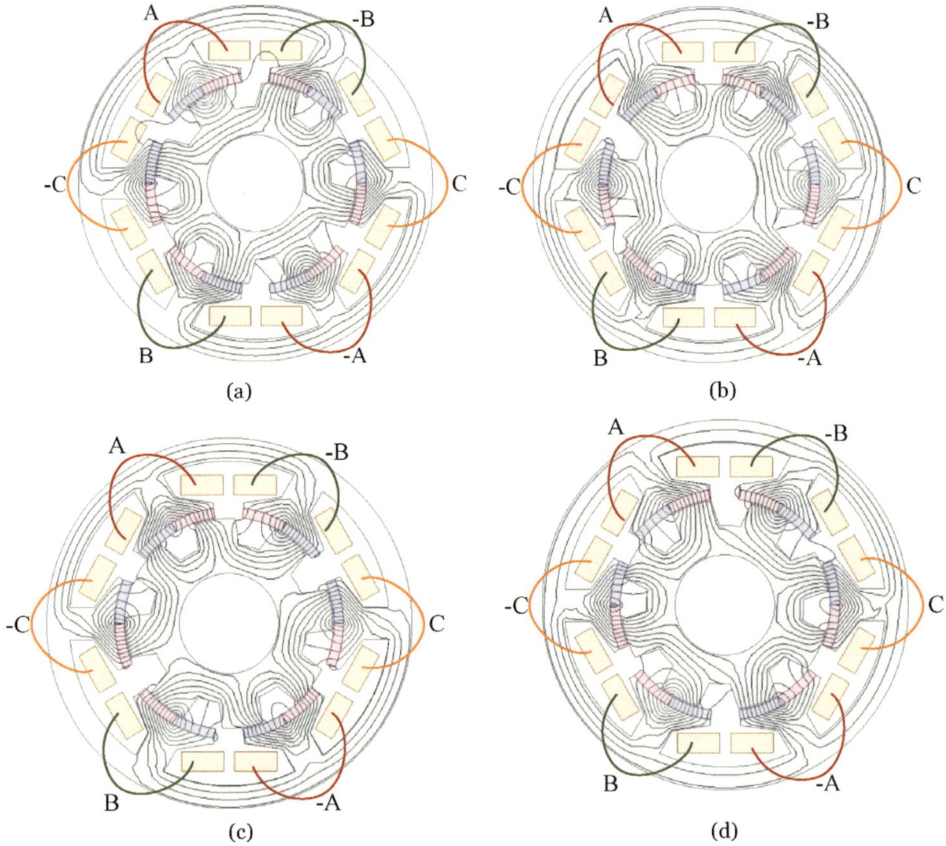

Figure 2.
No-load flux lines of the FRPMM excited by the PMs: (a) rotor position =0 elec. degree; (b) rotor position =90 elec. degree; (c) rotor position =180 elec. degree; (d) rotor position =270 elec. degree.

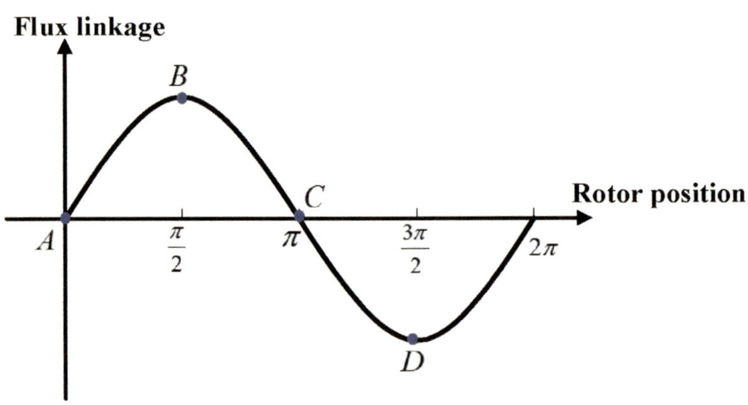

Figure 3.
Variation of flux linkage of phase a winding at different rotor positions.

after obtaining the bipolar flux linkage, as shown in **Figure 3**, the winding can produce a bipolar back-electromagnetic motive force (EMF). If the armature windings are injected with currents having the same frequency and phase with the back-EMF, a steady torque can be yielded.

3. Sizing equation of FRPMM

3.1 Magnetic circuit model

In order to derive the sizing equation of FRPMMs, the magnetic circuit model should be built at first; then, based on the model, the analytical equations of airgap flux density, back-EMF, and torque will be deduced.

The equivalent magnetic circuit model can be plotted as **Figure 4**. At No.1 stator tooth, its magnetic field distribution corresponds to the position shown in **Figure 2(b)**, that is, the rotor tooth is closer to the S-pole magnet. The S-pole magnetic generates two paths of magnetic flux, one is pole leakage flux Φ_{pl}, which goes through the adjacent N-pole magnet, the other is main flux Φ_m, which goes through the stator tooth, stator yoke, rotor tooth, and rotor yoke, thus can provide winding flux linkage and back-EMF. At No. 2 stator tooth, its magnetic field distribution corresponds to the position shown in **Figure 2(c)**, that is, the rotor axis is at the same distance from the S-pole and N-pole magnets. Thus, at this time, the two magnets can only generate one magnetic flux path, that is, the pole leakage flux Φ_{pl}. At No. 3 stator tooth, its magnetic field distribution corresponds to the position shown in **Figure 2(d)**, that is, the rotor tooth is closer to the N-pole magnet. The N-pole magnetic generates two paths of magnetic flux, one is pole leakage flux Φ_{pl}, which goes through the adjacent S-pole magnet, the other is main flux Φ_m, which goes through the stator tooth, stator yoke, rotor tooth, and rotor yoke, thus can provide winding flux linkage and back-EMF. It should be noted that the magnetic flux path of No. 1 stator tooth is just opposite to that of No. 3 stator tooth, so winding flux polarity in these two cases is just opposite to each other.

As mentioned above, **Figure 4** provides the magnetic circuit of FRPMMs, which can help analyze the flux distribution of FRPMMs at different rotor positions. However, the magnetic circuit requires the establishment of the whole FRPMM

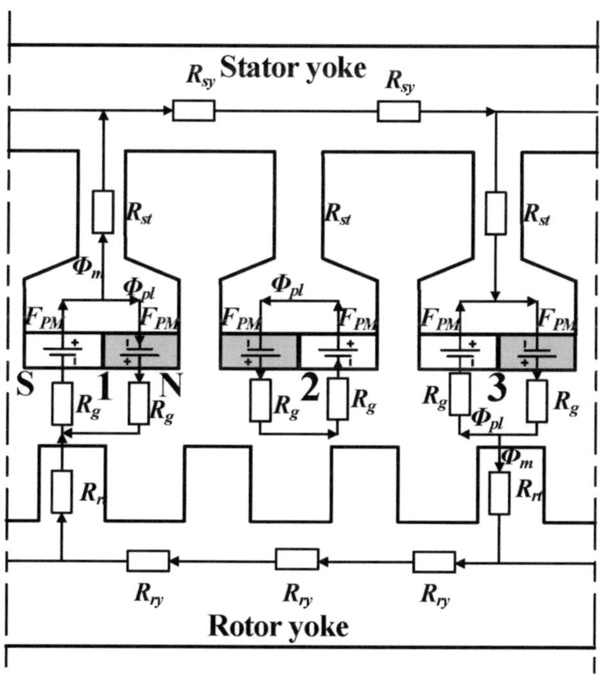

Figure 4.
Equivalent magnetic model of FRPMMs.

magnetic path, which is rather complex. Besides, the pole leakage flux, main flux, and the reluctance at each rotor positions should be calculated, which needs high workload. Therefore, a simplified magnetic circuit should be built. Observing **Figure 2**, it can be seen that a small rotor displacement brings a large rotation in stator flux field. This phenomenon is called as flux modulation effect, i.e. a high-pole slow-speed magnetic field becomes a low-pole high-speed magnetic field through the modulation effect of iron teeth. Therefore, the physical nature of FRPMM is indeed the flux modulation effect. The research of some flux modulation machines are usually based on the PM magnetic motive force (MMF)-airgap permeance model, such as the Vernier machine in [27]. So, this chapter will use this model to analyze FRPMMs.

In PM MMF-airgap permeance model, the no-load airgap flux density $B(\theta_s,\theta)$ can be written as the product of PM MMF $F_{PM}(\theta_s)$ and specific airgap permeance $\Lambda(\theta_s,\theta)$:

$$B(\theta_s,\theta) = F_{PM}(\theta_s)\Lambda(\theta_s,\theta) \tag{1}$$

where the definitions of angles θ_s and θ are shown in **Figure 5**. Then, the simplified magnetic circuit model can be given in **Figure 6**. Once knowing the PM

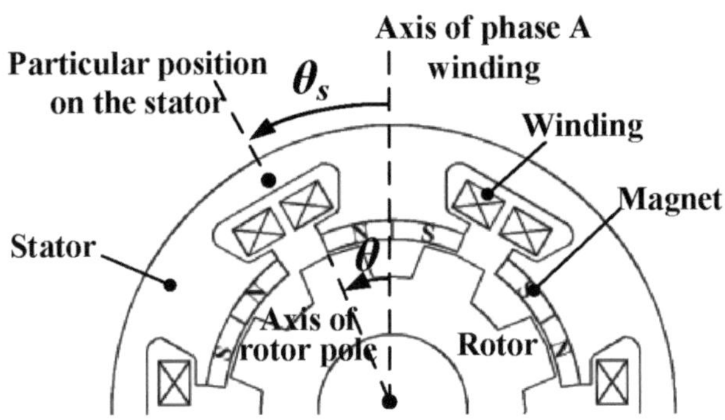

Figure 5.
Definitions of different angles in FRPMM.

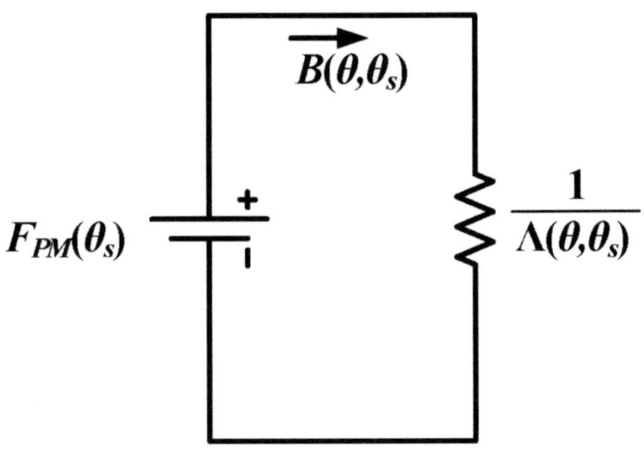

Figure 6.
Simplified equivalent magnetic model of FRPMMs.

MMF and airgap permeance, the no-load airgap flux density can be obtained. Then, the stator flux linkage $\lambda_{ph}(\theta)$ can be deduced using winding function theory:

$$\lambda_{ph}(\theta) = r_g l_{stk} \int_0^{2\pi} B(\theta_s, \theta) N(\theta_s) d\theta_s \tag{2}$$

where $N(\theta_s)$ is the phase winding function. After that, the phase back-EMF $E_{ph}(t)$ and average torque T_e can be calculated as:

$$E_{ph}(t) = \frac{d\lambda_{ph}(\theta)}{dt} \tag{3}$$

$$T_e = \frac{3}{2} E_{ph} I_{ph} \tag{4}$$

where I_{ph} is the peak value of phase current. Therefore, from Eqs. (1–4), it can be found that if the torque equation need to be calculated, the key is to obtain the equation of airgap flux density $B(\theta_s,\theta)$, which is further determined by the PM MMF $F_{PM}(\theta_s)$ and specific airgap permeance $\Lambda(\theta_s,\theta)$. Therefore, in the next parts, the equations of the PM MMF $F_{PM}(\theta_s)$ and specific airgap permeance $\Lambda(\theta_s,\theta)$ will be deduced in detail.

3.2 Airgap flux density equation

As aforementioned, to derive the torque equation, the no-load airgap flux density $B(\theta_s,\theta)$ should firstly be known, whose equation can be given as Eq. (1). Then, the next step is to derive the expressions of $F_{PM}(\theta_s)$ and $\Lambda(\theta_s,\theta)$. The PM MMF waveform excited by the magnets is shown in **Figure 7**, which can be given as:

$$F_{PM}(\theta_s) = \begin{cases} F_C; & 0 \le \theta_s < (1-SO)\pi/Z_s \\ 0; & (1-SO)\pi/Z_s \le \theta_s < (1+SO)\pi/Z_s \\ F_C; & (1+SO)\pi/Z_s \le \theta_s < 2\pi/Z_s \\ -F_C; & 2\pi/Z_s \le \theta_s < (3-SO)\pi/Z_s \\ 0; & (3-SO)\pi/Z_s \le \theta_s < (3+SO)\pi/Z_s \\ -F_C; & (3+SO)\pi/Z_s \le \theta_s < 4\pi/Z_s \end{cases} \tag{5}$$

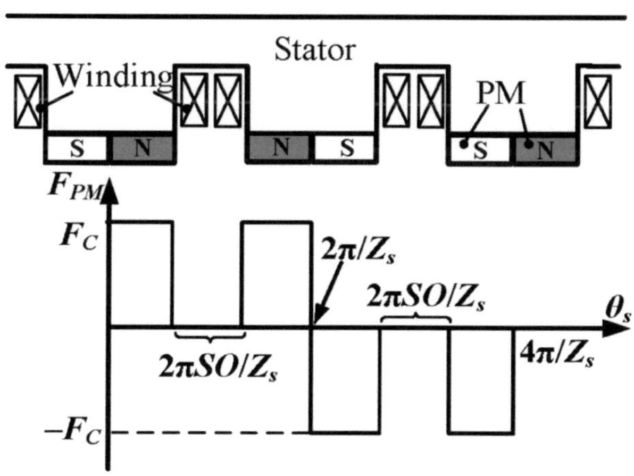

Figure 7.
Magnet MMF waveform.

where F_C is:

$$F_C = \frac{B_r h_m}{\mu_r \mu_0} \tag{6}$$

Then, it can be written in Fourier series as follows:

$$F_{PM}(\theta_s) = \sum_{i=1,3,5}^{\infty} F_i \sin\left(\frac{iZ_s}{2}\theta_s\right) \tag{7}$$

where the magnitude F_i is

$$F_i = \frac{4}{\pi}\frac{1}{i}\frac{B_r h_m}{\mu_0 \mu_r}\left[1 + (-1)^{\frac{i+1}{2}} \sin\left(\frac{i\pi}{2}SO\right)\right] \tag{8}$$

Then, the next step is to derive the specific airgap permeance $\Lambda(\theta_s,\theta)$ in Eq. (1). Since the stator slotting effect has already been considered in Eqs. (5–8), the specific airgap permeance $\Lambda(\theta_s,\theta)$ can be replaced by the airgap permeance with smoothed stator and slotted rotor $\Lambda_r(\theta_s,\theta)$. The model of smoothed stator and slotted rotor is shown in **Figure 8**. Then, the $\Lambda_r(\theta_s,\theta)$ can be expressed by:

$$\Lambda_r(\theta_s, \theta) \approx \Lambda_{0r} + \Lambda_{1r} \cos\left[Z_r(\theta_s - \theta)\right] \tag{9}$$

The coefficients of the airgap permeance function Λ_{0r} and Λ_{1r} in Eq. (9) can be obtained using the conformal mapping method [28, 29]:

$$\Lambda_{0r} = \frac{\mu_0}{g'}\left(1 - 1.6\beta\frac{b_o}{t}\right) \tag{10}$$

$$g' = g + h_m/\mu_r \tag{11}$$

$$\Lambda_{1r} = \frac{\mu_0}{g'}\frac{4}{\pi}\beta\left[0.5 + \frac{(b_o/t)^2}{0.78125 - 2(b_o/t)^2}\right]\sin\left(1.6\pi\frac{b_o}{t}\right) \tag{12}$$

$$\beta = 0.5 - \frac{1}{2\sqrt{1 + \left(\frac{b_o}{2t}\frac{t}{g'}\right)^2}} \tag{13}$$

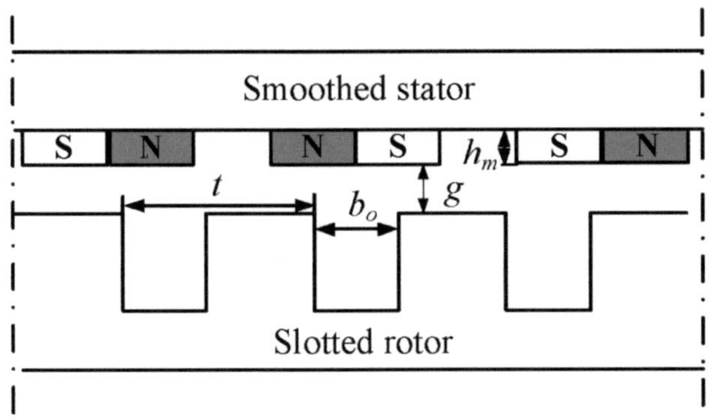

Figure 8.
Schematic of single-side salient structure on rotor.

where b_o is the rotor slot opening width and t is the rotor slot pitch, as shown in **Figure 8**. Combining Eq. (1), Eqs. (5–13), the no-load airgap flux density $B(\theta_s,\theta)$ can be finally calculated as:

$$B(\theta_s, \theta) = \sum_{i=1,3}^{\infty} B_i \sin\left[\left(\frac{iZ_s}{2} \pm Z_r\right)\theta_s - Z_r\theta\right] \tag{14}$$

where the magnitude B_i is

$$B_i = \frac{1}{2} F_i \Lambda_{1r},\ i = 1, 3, 5 \ldots \tag{15}$$

3.3 Slot-pole combinations

As can be seen in Eq. (14), the number of pole pairs in the air gap flux density is $iZ_s/2 \pm Z_r$, i = 1,3,5 ... Then, in order to make the flux density induce EMF in the armature windings, the pole pair number of the armature windings P should be equal to $iZ_s/2 \pm Z_r$, i = 1,3,5 ... Besides, for three phase symmetry, the winding pole pair number must also meet the following requirement:

$$\frac{Z_s}{GCD(Z_s, P)} = 3k,\ k = 1, 2, 3 \ldots \tag{16}$$

All in all, the slot-pole combination of three-phase FRPMMs is ruled by the following equation:

$$\begin{aligned} P = \min &\left\{ P = \frac{iZ_s}{2} \pm Z_r;\quad \frac{Z_s}{GCD(Z_s, P)} = 3k \right\} \\ i = 1, 3, 5 \ldots &\quad k = 1, 2, 3 \ldots \end{aligned} \tag{17}$$

where min means to select the minimum number of these qualified harmonic orders so as to obtain a maximal pole ratio of FRPMMs. Therefore, the feasible slot-pole combinations can be summarized as **Table 2**. Non-overlapping windings (i.e., concentrated windings) are usually used in FRPMMs because of the higher fault tolerance and easier manufacture than regular overlapping windings. However, some FRPMMs are suggested to employ overlapping windings in order to have a larger winding factor and thus a higher torque density. Therefore, both winding factors, that is, k_{wn} (using non-overlapping winding) and k_{wr} (using overlapping winding) are calculated for each FRPMM so as to see the difference of using different winding types.

3.4 Torque equation

Once the stator winding pole pair is selected, the stator flux linkage can be deduced using winding function theory, just as mentioned in Eq. (2). The winding function $N(\theta_s)$ in Eq. (2) can be written as:

$$N(\theta_s) = \sum_{i=1,3,5}^{\infty} N_i \cos(iP\theta_s) \tag{18}$$

$$N_i = \frac{2}{i\pi}\frac{N_s}{P} k_{wi} \tag{19}$$

Z_s	Z_r	2	3	4	5	6	7	8	10	11	12	13	14	15	16
6	*P*	1		1	2		2	1	1	2		2	1		1
	SPP	1		1	0.5		0.5	1	1	0.5		0.5	1		1
	PR	2		4	2.5		3.5	8	10	5.5		6.5	14		16
	k_{wn}	0.5		0.5	0.866		0.866	0.5	0.5	0.866		0.866	0.5		0.5
	k_{wr}	1		1	0.866		0.866	1	1	0.866		0.866	1		1
12	*P*	4		2	1		1	2	4	5		5	4		2
	SPP	0.5		1	2		2	1	0.5	0.4		0.4	0.5		1
	PR	0.5		2	5		7	4	2.5	2.2		2.6	3.5		8
	k_{wn}	0.866		0.5	0.25		0.25	0.5	0.866	0.933		0.933	0.866		0.5
	k_{wr}	0.866		1	0.966		0.966	1	0.866	0.933		0.933	0.866		1
18	*P*	7	6	5	4	3	2	1	1	2	3	4	5	6	7
	SPP	3/7	0.5	0.6	0.75	1	1.5	3	3	1.5	1	0.75	0.6	0.5	3/7
	PR	2/7	0.5	0.8	1.25	2	3.5	8	10	5.5	4	3.25	2.8	2.5	16/7
	k_{wn}	0.902	0.866	0.735	0.617	0.5	0.492	0.167	0.167	0.492	0.5	0.617	0.735	0.866	0.902
	k_{wr}	0.902	0.866	0.945	0.945	1	0.945	0.96	0.96	0.945	1	0.945	0.945	0.866	0.902
PS:	Non-overlapping winding is recommended.									Other: Overlapping winding is recommended.					

k_{wn} *and* k_{wr} *are fundamental winding factors calculated based on non-overlapping winding type and recommended winding types, respectively.*

Table 2.
Slot-pole combinations of three-phase FRPMM.

where N_i is the ith harmonics of the winding function and k_{wi} is the winding factor of the ith harmonics. As can be seen in Eq. (17), the pole pair number is $iZ_s/2 \pm Z_r$ (i = 1,3,5 ...). So, the sum or difference of any two pole pair harmonics P_{i1} and P_{i2} is a multiple of stator slot number, that is,

$$\begin{cases} P_{i1} = i_1 Z_s/2 \pm Z_r \\ P_{i2} = i_2 Z_s/2 \pm Z_r \\ |P_{i1} \pm P_{i2}| = kZ_s, \quad k = 1, 2, 3 \ldots \end{cases} \tag{20}$$

Therefore, all the flux density harmonics are tooth harmonics of each other, that is, they have the same absolute values of winding factors, and their absolute winding factor equals the fundamental winding factor k_{w1}:

$$|k_{wP_{i1}}| = |k_{wP_{i2}}| = k_{w1} \tag{21}$$

Then, combining Eq. (2), Eq. (3), Eqs. (18–21), the back-EMF can be finally obtained as:

$$E_{ph} = 2\omega_m r_g l_{stk} N_s Z_r k_{w1} \sum_{i=1}^{\infty} \text{sgn} * \frac{B_i}{(\frac{iZ_s}{2} \pm Z_r)/P} \tag{22}$$

where

$$\text{sgn} = \begin{cases} 1, \text{winding factor of } (iZ_s/2 \pm Z_r)^{\text{th}} \text{ harmonic equals } k_{w1} \\ -1, \text{winding factor of } (iZ_s/2 \pm Z_r)^{\text{th}} \text{ harmonic equals } - k_{w1} \end{cases} \tag{23}$$

Since the reluctance torque of FRPMM is negligible, the electromagnetic torque under i_d = 0 control can be expressed as Eq. (4). Then, combining Eq. (4) and Eq. (22), the average torque T_e is able to be calculated as:

$$T_e = 3I_{ph} r_g l_{stk} N_s Z_r k_{w1} \sum_{i=1}^{\infty} \text{sgn} * \frac{B_i}{(\frac{iZ_s}{2} \pm Z_r)/P} \tag{24}$$

So far, the general torque equation has been obtained as Eq. (24), but in this equation, some parameters such as B_i, I_{ph} cannot be determined in the initial design stage of FRPMMs, so it is desirable that Eq. (24) can be transformed to a combination of several basic parameters, such as electric loading, magnetic loading, which can be easily determined in the initial design stage.

As known for electrical machines, the electric loading A_e can be written as:

$$A_e = \frac{6N_s I_{ph}}{2\sqrt{2}\pi r_g} \tag{25}$$

Then, the equivalent magnetic loading of three-phase FRPMM B_m is defined as:

$$B_m = \sum_{i=1}^{\infty} \text{sgn} * \frac{B_i}{(\frac{iZ_s}{2} \pm Z_r)/P} \tag{26}$$

So, the torque expression in Eq. (24) can be rewritten as:

$$T_e = \sqrt{2}\pi r_g^2 l_{stk} k_w Z_r A_e B_m \tag{27}$$

Thus, the rotor volume V_r, which equals $\pi l_{stk} r^2 g$, can be obtained:

$$V_r = \frac{T_e}{\sqrt{2} k_w Z_r A_e B_m} \tag{28}$$

and then the airgap radius r_g and the stack length l_{stk} can be derived as:

$$r_g = \sqrt[3]{V_r/(\pi k_{lr})} \tag{29}$$

$$l_{stk} = \sqrt[3]{V_r k_{lr}^2/\pi} \tag{30}$$

where k_{lr} is the aspect ratio, equals to the ratio of r_g to l_{stk}. It can be found in Eq. (27) that the key parameters affecting the torque density are the airgap radius r_g, stack length l_{stk}, winding factor k_w, rotor slot number Z_r, electric loading A_e, and equivalent magnetic loading B_m, among which the stack length l_{stk} can be determined by the volume requirement, and winding factor k_w is approximate to 1. So, the remaining parameters r_g, Z_r, A_e, B_m should be determined at the initial stage of the design process. Thus, the influences of the above key parameters on important performances, such as average torque, pulsating torque, power factor, PM demagnetization performance, will be investigated in the following parts.

4. Influence of design parameters on key performances

4.1 Average torque performances

4.1.1 Influence of slot-pole combinations on average torque

As aforementioned, the rotor slot number Z_r is one of key parameters that should be determined in the first design stage. How to determine the rotor slot number is a question. In this part, the influence of Z_r on the torque performance will be investigated, giving instruction on how to select Z_r. The parameters of the FRPMM models are listed in **Table 3**. These parameters are kept the same for the FRPMMs in order to have a reasonable comparison of their torque performance. That is to say, the airgap radius r_g, stack length l_{stk}, and electric loading A_e are the same.

Figure 9 shows the influence of rotor slot number on the output torque when non-overlapping windings and recommended windings are used respectively. For **Figure 9(a)**, when non-overlapping windings are adopted, the average torque is mainly related to the product of winding factor and rotor slot number, that is,

Parameter	Value	Parameter	Value
Stator outer diameter	170 mm	Stator inner diameter	105 m
Stator slot opening ratio	0.25	Remanent permeability	1.065
Stack length	100 mm	PM thickness	2.5 mm
Series turns per phase	80	Airgap length	0.5 mm
Rotor slot opening ratio	0.65	Rated current	5.3A
Rated speed	600 rpm	Magnet remanence	1.21 T

Table 3.
Parameters of the three-phase FRPMM models.

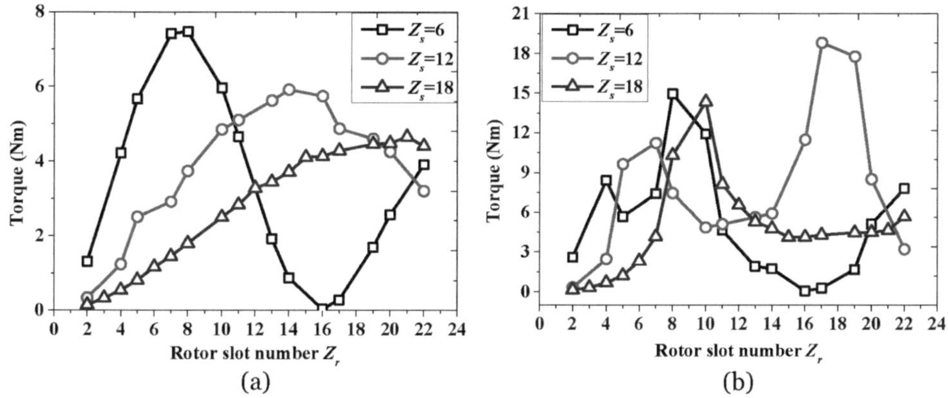

Figure 9.
Effect of combinations of stator slots and rotor slots on torque: (a) non-overlapping windings; (b) recommended windings.

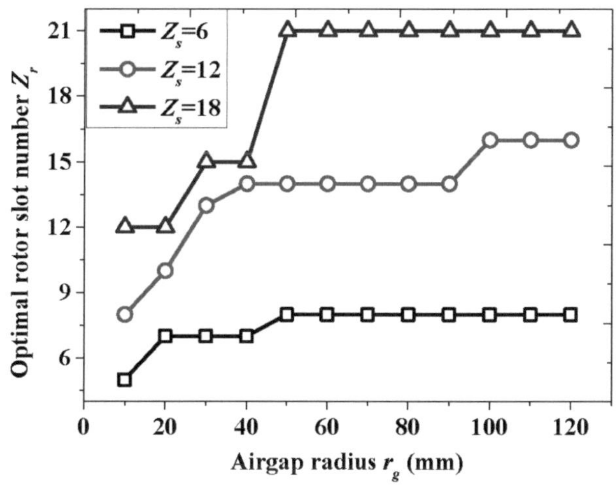

Figure 10.
Effect of r_g *on optimal* Z_r *when split ratio is 0.6.*

$k_w{}^*Z_r{}^*B_m$. Since the machine volume and PM usage are kept the same, the equivalent magnet loading B_m is mainly determined by the pole ratio (PR). So, the variation trend of torque is similar to that of $k_w{}^*Z_r{}^*$PR. It can be seen that the torque achieves the maximum value when the rotor slot number is 8, 14, and 21 for 6 stator slots, 12 stator slots, and 18 stator slots, respectively. When the recommend windings are used, which means that the winding factor are maximized, the main factor that affects the torque is the $Z_r{}^*$PR. As shown in **Table 2**, the variation of PR is irregular, hence the variation of torque with rotor slot number is irregular. As can be seen, for recommended winding types, the torque achieves the maximal value when the rotor slot number is 8, 10 and 17 for 6 stator slots, 12 stator slots, and 18 stator slots, respectively.

4.1.2 Influence of airgap radius on average torque

As shown in Eq. (27), the airgap radius r_g is also very important for the output torque. **Figure 10** investigates the effect of optimal rotor slot number Z_r at different r_g. For 6, 12, 18 stator slots, their rotor slot numbers are selected as 8, 14, and 21, respectively. Moreover, non-overlapping windings are used in these models because non-overlapping winding is simple and has the same end winding length. It can be seen that when the airgap radius is small, the optimal rotor slot number is small. This is because when the airgap radius is small, the leakage flux between adjacent rotor teeth occupies a large percent, so the optimal rotor slot number should be small to reduce the leakage flux as much as possible. When the airgap radius gets larger and larger, the leakage flux decreases gradually. Hence, the optimal rotor slot number increases.

Then, keeping the stator outer diameter as a constant, that is, 170 mm, the effects of airgap radius of average torque are analyzed in **Figure 11**. It indicates that when the airgap radius increases, the output torque goes up. This is because the torque is proportional to the square of airgap radius. The larger the airgap radius, the higher the torque. However, the torque is not only influenced by the airgap radius, but also the electric loading A_e. With the increase of airgap radius, the inner diameter of the stator increases, and thus the slot area decreases, leading to the decrease of winding turns per slot and the electric loading. Therefore, as the airgap radius keeps increasing, the output torque decreases afterwards.

4.1.3 Influence of magnetic loading and equivalent electric loading on average torque

In addition to the rotor slot number Z_r, airgap radius r_g, the rest of key parameters affecting the torque in Eq. (27) are the electric loading A_e and the equivalent magnetic loading B_m. **Figure 12** analyzes the influence of A_e and B_m on the average torque at different stator slot number. For these models, the airgap radius is fixed as 55 mm and their rotor slot number is chosen as their corresponding optimal value. Also, non-overlapping windings are adopted. As can be seen, the output torque increases with the electric loading. This reason is very simple, that is, a larger current, a higher torque. But for the equivalent magnetic loading, the variation trend of torque does not monotonically increase with the equivalent magnetic loading. This is due to the saturation effect of the iron core. Moreover, it can be seen that the knee point of the equivalent magnet loading increases with the stator slot number. Since the winding pole pair of the 18-stator-slot FRPMM is 6, which is larger than 1-winding-pole-pair of the 6-stator-slot and 4-winding-pole-pair of the

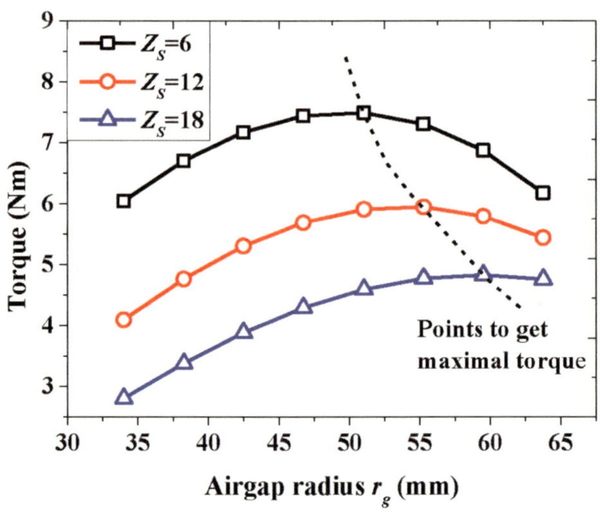

Figure 11.
Effect of r_g *on torque when stator outer diameter is 170 mm.*

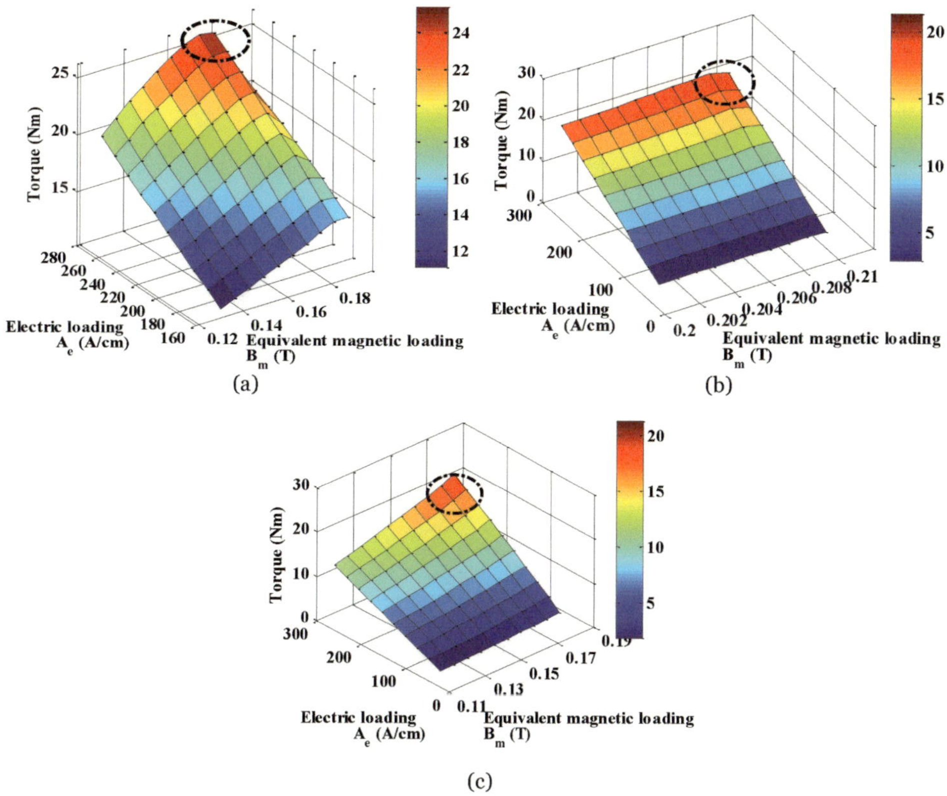

Figure 12.
Effect of equivalent magnetic loading and electric loading on torque: (a) Z_s = 6; (b) Z_s = 12; (c) Z_s = 18.

12-stator-slot FRPMM, the stator iron of the 18-stator-slot FRPMM is less likely to saturate than the others.

4.2 Pulsating torque performances

4.2.1 Influence of slot-pole combinations on pulsating torque

Apart from the torque density, pulsating torque is also very important because a large pulsating torque will increase the vibration and noise of machines. **Figure 13** shows the cogging torque and ripple torque waveforms of 13-, 14-, 16-, 17-, and 19-rotor-slot FRPMMs. The stator slot number of these models is all chosen as 12. For the rated torque, we can see in **Figure 13(b)** that the 14-rotor-slot FRPMM yields the largest among the five models. As for the pulsating torque, we can see that the cogging torque and ripple torque of 16-rotor-slot FRPMM are the largest, and that of 19-rotor-slot FRPMM is the least. This phenomenon is related to the least common multiple of stator slot number and rotor slot number. The larger least common multiple, the lower pulsating torque. The least common multiples of the 13-, 14-, 16-, 17-, and 19-rotor-slot FRPMMs are 156, 84, 48, 204, and 228, respectively. Therefore, the 19-rotor-slot FRPMM exhibit the lowest cogging torque and ripple torque. However, attentions should be paid to use odd rotor number because it will cause other problems such as eccentricity stress. **Figure 14** compares the radial stress of the five FRPMM models. It can be seen that for the even rotor slot number FRPMMs, that is, 14 and 16 rotor slots, the stress harmonics only have even orders, which will not lead to eccentricity. However, for the odd rotor slot number

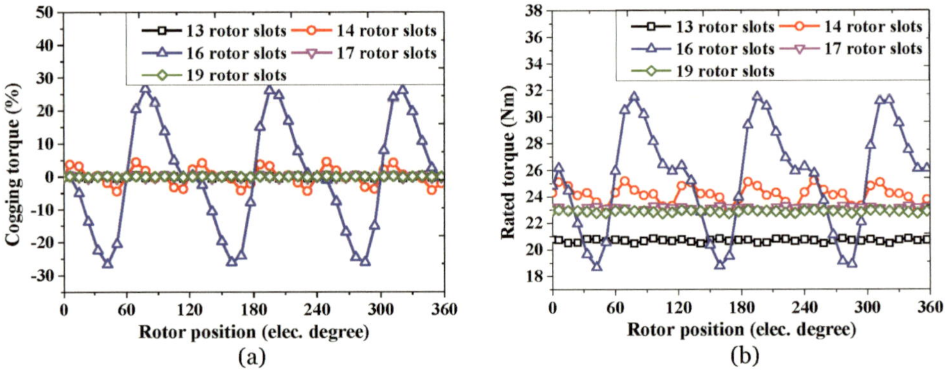

Figure 13.
Effect of slot-pole combination on pulsating torque performances: (a) cogging torque waveforms (%); (b) rated torque waveforms.

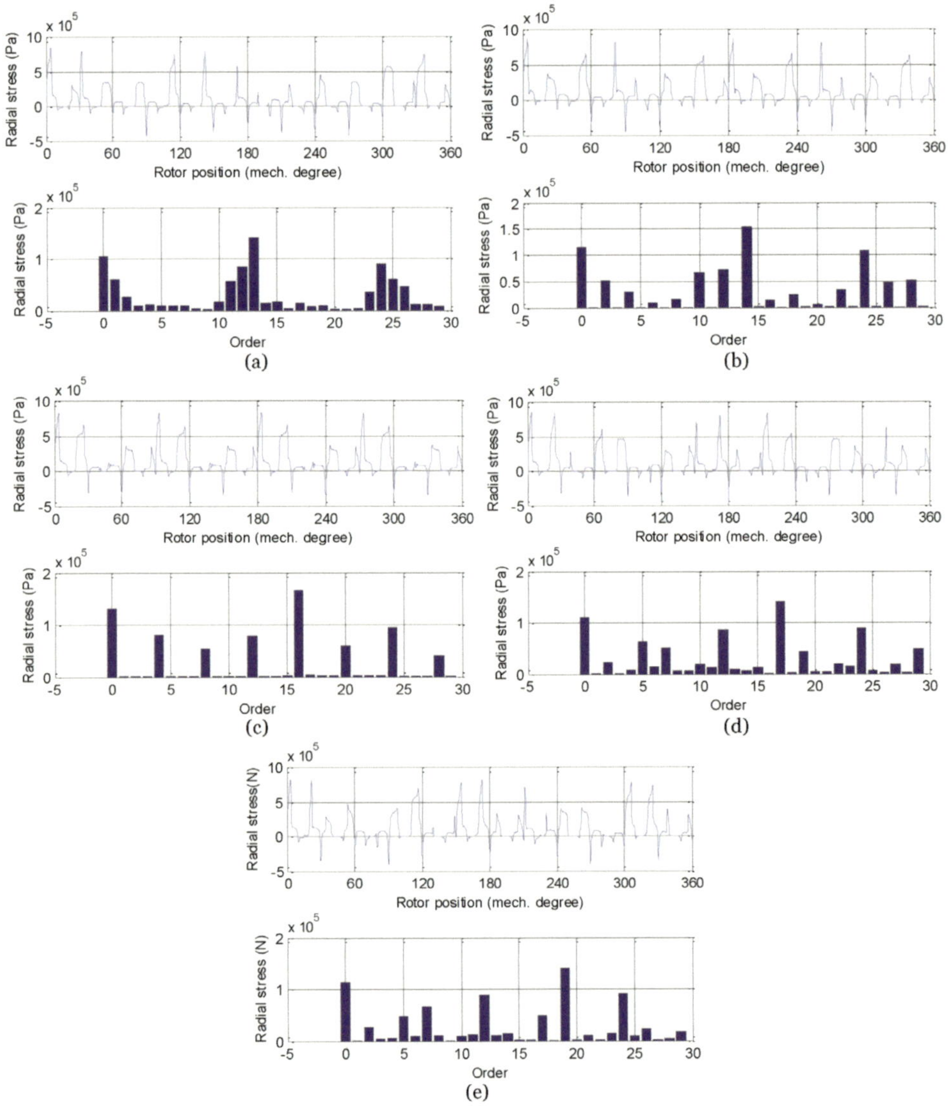

Figure 14.
Radial stress analysis of the FRPMMs: (a) 13-rotor-slot; (b) 14-rotor-slot; (c) 16-rotor-slot; (d) 17-rotor-slot; (e) 19-rotor-slot.

FRPMMs, that is, 13, 17, and 19 rotor slots, there are many odd stress harmonics. Since the first-order harmonic is dominant for the eccentricity, the 13-rotor-slot FRPMM has a large eccentricity stress. Therefore, 13-rotor-slot is not recommended. The first-order stress harmonic for 17 and 19 rotor slots are very small, so their eccentricity can be neglected.

4.2.2 Influence of PM thickness and split ratio on pulsating torque

The influences of split ratio and PM thickness on cogging torque and ripple torque of FRPMMs are also analyzed in **Figure 15**. This figure is plotted based on the 14-rotor-slot, which is chosen because it has the largest torque density and a relatively low pulsating torque, as shown in **Figure 13**. It can be found in **Figure 15(a)** that the cogging torque increases with the PM thickness and the split ratio. When the PM thickness increases, the airgap flux density increases, and thus the interaction between the PMs and slot-teeth becomes greater, which leads to a higher cogging torque. As the split ratio increases, the airgap radius increases, hence the cogging torque increases with the split ratio [30]. As for the ripple torque, the ripple torque has the maximum value when the split ratio is around 0.66. This is because the ripple torque is not only related to the slot structure but also influenced by the electric loading. As aforementioned, the pulsating torque resulting from the slot structure is increased with the split ratio. However, as the split ratio increases, the slot area is reduced and the electric loading gets smaller and smaller, so the ripple torque resulting from the electric loading becomes lower. Considering these two impacts, the ripple torque has a maximal value when the split ratio changes.

4.2.3 Influence of slot opening ratios on pulsating torque

As we know, the airgap structure is significant for the pulsating torque because the pulsating torque results from the interaction between the two sides of the airgap, that is, stator and rotor. Therefore, this chapter also analyzes the influences of stator slot opening ratio and rotor slot opening ratio on cogging torque and ripple torque. Here, the stator/rotor slot opening ratio is defined as the ratio of stator/rotor slot opening width to the stator/rotor slot pitch. **Figure 16** shows the variation of cogging torque and ripple torque with the two slot opening ratios. It can be seen that the cogging torque increases with the stator slot opening ratio. The reason is that a larger stator slot opening ratio reduces the PM width and the smoothness of PM MMF, thus the changing of the PM MMF along the tangential direction increases

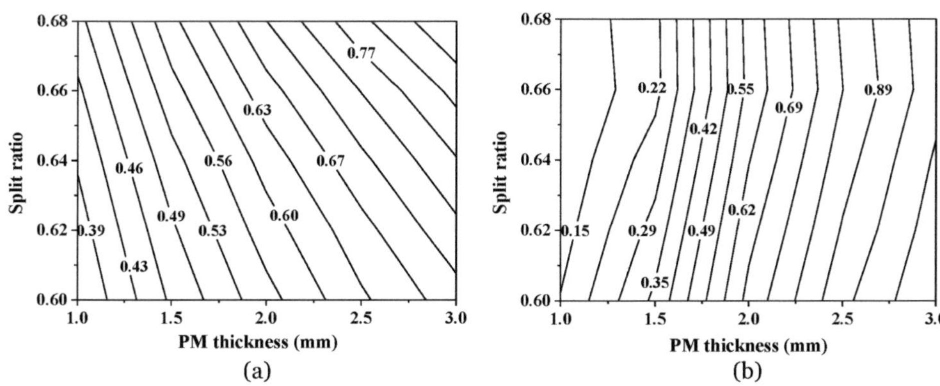

Figure 15.
Effect of split ratio and PM thickness on pulsating torque performances: (a) cogging torque (%); (b) ripple torque (%).

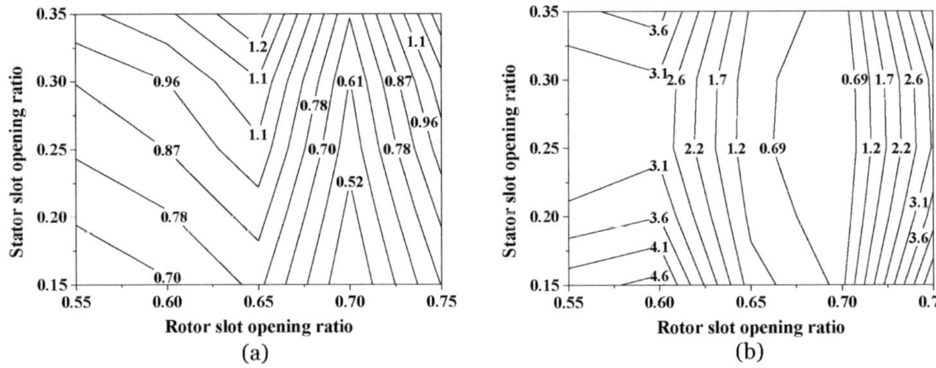

Figure 16.
Effect of stator slot opening ratio and rotor slot opening ratio on pulsating torque: (a) cogging torque (%); (b) ripple torque (%).

the cogging torque. As for the rotor slot opening ratio, which simultaneously influences all the harmonic contents of the airgap permeance, it has great and nonlinear impact on the pulsating torque. Since the pulsating torque results from the interaction of multi permeance harmonics, the variation of pulsating torque changes nonlinearly with the rotor slot opening ratio. It can be seen in **Figure 16** that the optimal cogging torque and ripple torque can be achieved when the stator slot opening ratio and rotor slot opening ratio are around 0.25 and 0.7, respectively.

4.3 Power factor performances

4.3.1 Influence of stator inner diameter and PM thickness on power factor

Since the power factor of FRPMMs is usually low, which is around 0.4–0.7, meanwhile a low power factor will increase the converter capacity and cost, the influences of key parameters on the power factor should be also analyzed to achieve a relatively high power factor. The power factor can be given as:

$$PF = 1/\sqrt{1 + \left(\frac{L_s I_s}{\psi_m}\right)^2} \tag{31}$$

where I_s is the winding current, L_s is the synchronous inductance (because the saliency ratio is approximate to 1, $L_d \approx L_q$), and ψ_m is the PM flux linkage. Then, the effect of stator inner diameter on power factor is shown in **Figure 17**. Here, the stator outer diameter is kept as 124 mm, and the airgap length is fixed as 0.5 mm. It can be found that with the increase of stator inner diameter, the power factor increases continuously. The reason is that with the increase of stator inner diameter, the slot area decreases, so the winding turns per phase decreases, thus leading to the reduction of the synchronous inductance L_s. The lower L_s, the higher power factor, as shown in Eq. (31). Apart from the stator inner diameter, another important parameter affecting the power factor is the PM thickness h_m. **Figure 18** investigates the variation of power factor with respect to the PM thickness. It indicates that the power factor initially increases with the PM thickness but then decreases. The reason is explained as follows. As the PM thickness increases, the PM flux linkage ψ_m becomes larger, so the power factor increases. However, the synchronous inductance L_s also increases with the PM thickness, which leads to the reduction of power factor afterwards. Therefore, there is an optimal PM thickness for a maximum achievable power factor.

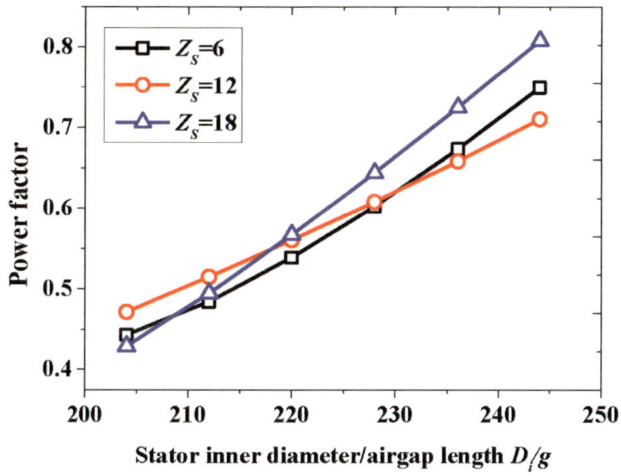

Figure 17.
Effect of stator inner diameter/airgap length on power factor.

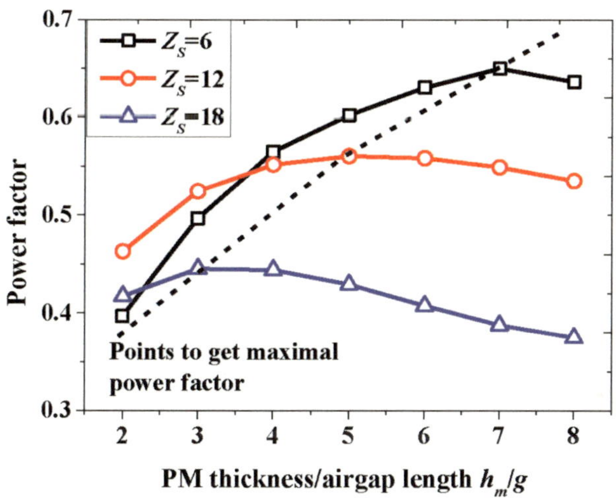

Figure 18.
Effect of PM thickness/airgap length on power factor.

4.3.2 Influence of slot opening ratio on power factor

Another important parameter that influences the airgap structure is the slot opening ratio. Hence, **Figures 19** and **20** analyzes the effect of stator slot opening ratio and rotor slot opening ratio on pulsating torque performances, respectively. It can be seen in **Figure 19** that the maximum power factor can be obtained when the stator slot opening ratio is approximately to 0.3. The explanation is as follows. When the stator slot opening ratio is too small, the slot leakage flux between the stator tips is large, thus the main flux is reduced, and the back-EMF is lowered, resulting in smaller back-EMF. And when the stator slot opening ratio is too large, the PM width will be narrower. Although the slot leakage flux is reduced, the main flux is not high due to the narrower PMs, thus the back-EMF is lowered. Therefore, the stator slot opening ratio cannot be too small or too large, that is, there is an optimal value for the stator slot opening ratio.

Then, the influences of rotor slot opening ratio on power factor can be seen in **Figure 20**. It indicates that when the rotor slot opening ratio is around 0.7, the

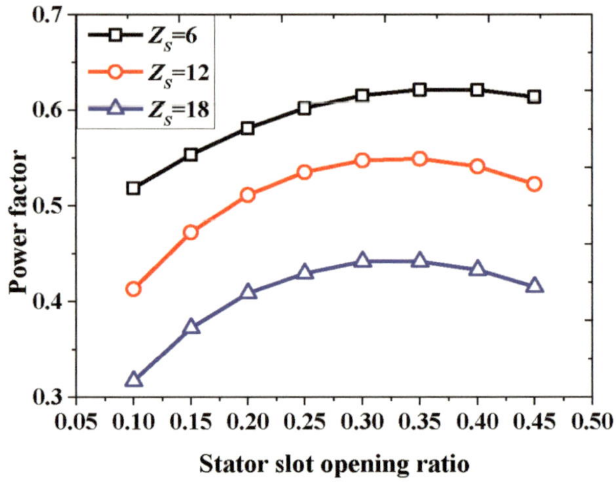

Figure 19.
Effect of stator slot opening ratio on power factor.

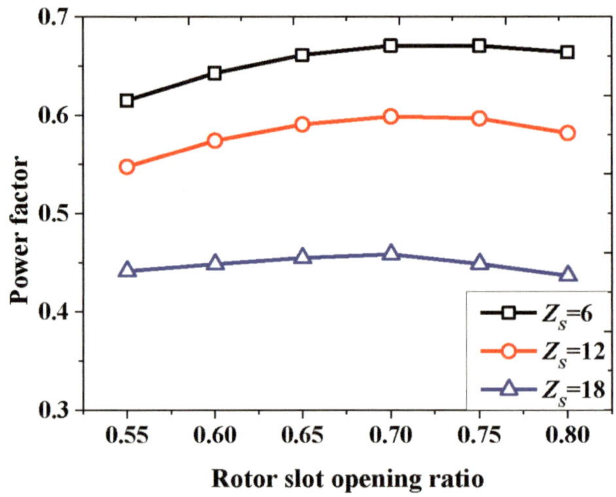

Figure 20.
Effect of rotor slot opening ratio on power factor.

power factor reaches the maximal value. This is because the power factor is mainly influenced by the back-EMF. When the rotor slot opening ratio increases, the effective airgap length becomes smaller, thus the main flux is increased and the back-EMF is improved. As a result, the power factor is increased. When the rotor slot opening ratio keeps increasing, the flux modulation effect of the rotor teeth becomes weaker and weaker, thus the smaller modulated flux, and the lower back-EMF. Therefore, there is also an optimal value for rotor slot opening ratio when a high power factor is demanded.

4.4 PM demagnetization performances

For PM machines, PM demagnetization performances are very important because it is highly related to the safe operation and machine reliability. Therefore, the PM demagnetization performances of FRPMMs should be analyzed in this chapter. Since the magnetic properties of PM materials are sensitive to temperature, and the temperature coefficient of NdFeB magnet is as high as $-0.126\%K^{-1}$. When the current of FRPMMs is large, the winding heating can easily affect the PMs attached to the stator teeth surface, causing the decrease of PM magnetic performances. On the other hand, when the winding current is large, the demagnetizing effect of the armature field is enhanced, and thus the PMs have the possibility to be demagnetized. Therefore, it is of great importance to investigate the PM demagnetization performances of FRPMMs at different conditions.

Figure 21 shows the demagnetization curve of the magnets. The upper half is a straight line, and lower half under the knee point B_{knee} is a curved line. When the FRPMM works on the straight line (such as point P_1), the return line coincides with the demagnetization curve, and the magnetic performance of the magnets will not be lost. However, when the armature equivalent MMF $H_a{}'$ is too large at load condition, or the knee point is too high, the working point B_{knee} is moved to P_2. At this time, the recovery line does not coincide with the original demagnetization line, thus the intersection of the B-axis changes from B_r to B_{r1}, causing the irreversible demagnetization. Then, the PM properties and machine performances will no longer return to the original. So, the PM flux density should be examined in order to check the risk of irreversible demagnetization. As we know, the PM flux density is determined by the design parameters such as electric loading A_e, PM thickness h_m, rotor slot opening ratio, etc. So, in this chapter, the effects of electric loading A_e, PM thickness h_m, rotor slot opening ratio b_o/t on PM demagnetization performances of FRPMMs will be studied. For instance, the PM material is selected as N38SH, and knee point of the PM flux density at 100°C is 0.35 T.

Figure 22 shows the PM flux density of a 12-stator-slot/14-rotor-slot FRPMM when the electric loading A_e is 1600A/cm, the PM thickness h_m is 3 mm, rotor slot opening ratio b_o/t is 0.65. It can be seen that the PM flux density distribution varies with the rotor position. When the rotor position is 140°, the PM does not demagnetize, while at 0° and 340°, the PM will demagnetize. Hence, in the following analysis, the PM flux density at the most severe moment of demagnetization is selected.

Figure 23 studies the magnetic flux density distribution in the PMs under different electric loadings. It can be found that the larger electric loading A_e, the smaller minimum flux density. This is because the larger electric loading, the higher armature MMF $H_a{}'$, and the more left operating point P_2, so the lower flux density

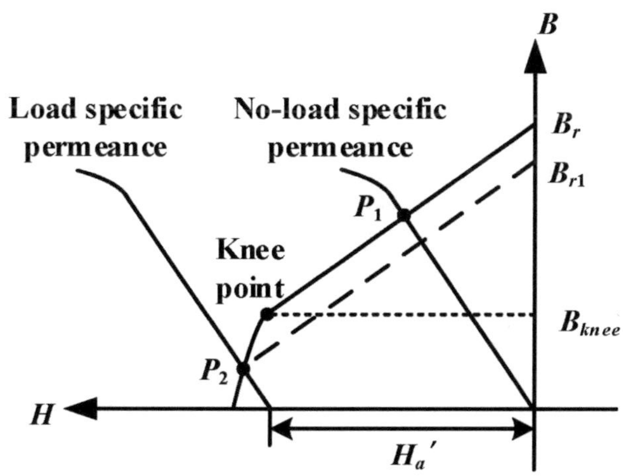

Figure 21.
PM demagnetization curve.

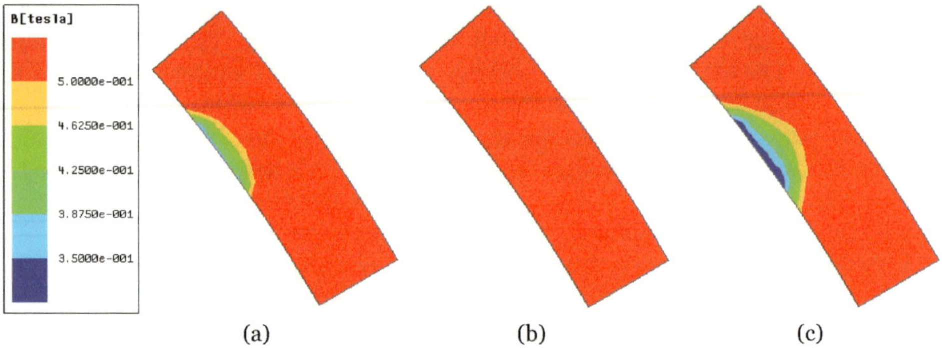

Figure 22.
PM demagnetization at different rotor positions: (a) rotor position = 0°; (b) rotor position = 140°; (c) rotor position = 340°.

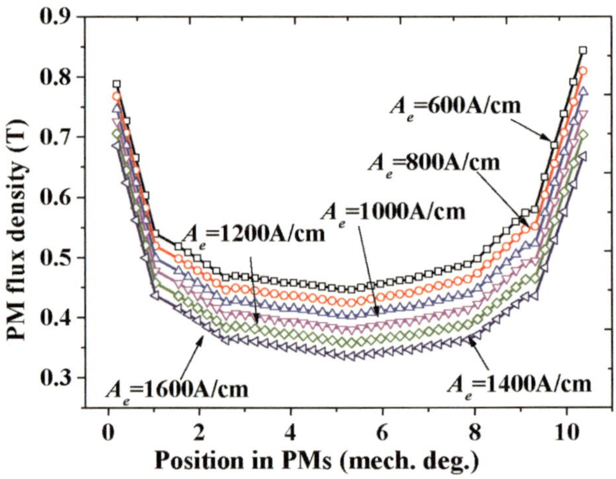

Figure 23.
Influence of A_e *on PM demagnetization.*

in the magnets. When the electric loading A_e is 1400A/cm, the PM irreversible demagnetization just occurs. In addition, it can be seen that the entire magnetic flux density map is skewed to the right. This is because the N-pole magnet is intercepted in this analysis, and there is an S-pole magnet next to the N-pole magnet. There is PM pole leakage flux between the S-pole magnet (negative axis) and the N-pole magnet (positive axis), so the magnetic flux density around the 0 position is lower, and away from the 0 position, the magnetic flux density gradually rises.

Figure 24 analyzes the effect of PM thickness h_m on the PM demagnetization performances. At this time, the electric loading is chosen as 800 A/cm, and the rotor slot opening ratio is selected as 0.65. It can be seen in **Figure 24** that when the PM thickness h_m is less than 2.5 mm, the irreversible demagnetization will happen, while when the PM thickness h_m is larger than 2.5 mm, the irreversible demagnetization will not. In this model, the airgap length is 0.5 mm. Therefore, in the design stage, the PM thickness should be better to set as five times or more the airgap length. Considering the back-EMF, it is claimed in [3] that when the PM thickness is about three times the airgap length, the back-EMF will reach the maximum. But considering both back-EMF and PM demagnetization risk, it is safer to set the PM thickness as about five times airgap length.

Figure 25 shows the influences of rotor slot opening ratio b_o/t on the flux density distribution inside the PMs. At this time, the electric loading is chosen as 800 A/cm,

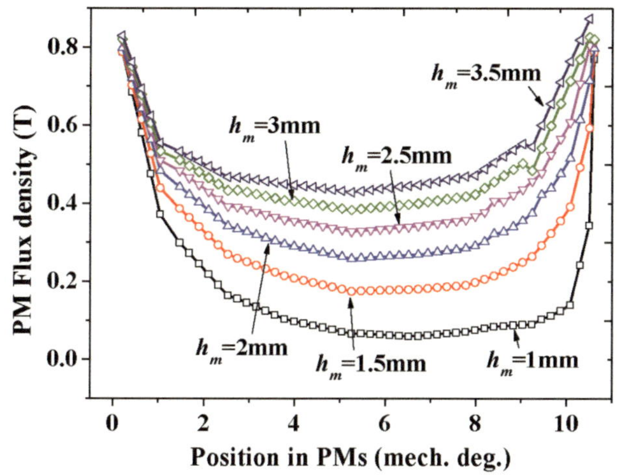

Figure 24.
Influence of h_m *on PM demagnetization.*

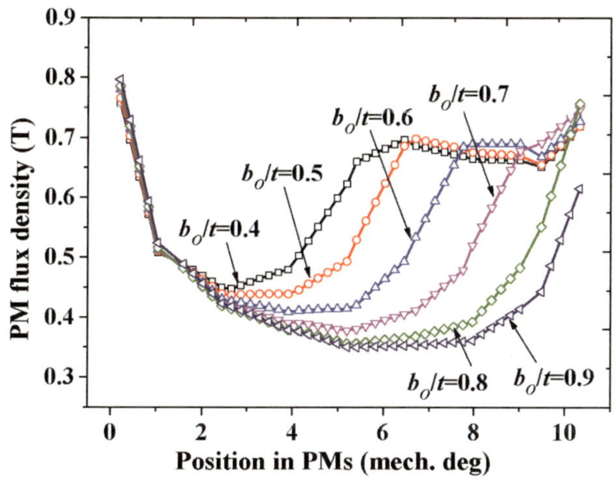

Figure 25.
Influence of b_o/t *on PM demagnetization.*

and the PM thickness is selected as five times the airgap length, that is, 3 mm. The larger rotor slot opening ratio, the narrower rotor teeth, thus the more saturated rotor teeth, and the smaller magnetic reluctance. As shown in **Figure 20**, when the magnetic gets smaller, the more left operating point P_2, and thus the lower PM flux density. It can be seen in **Figure 25** that when the rotor slot opening ratio b_o/t is 0.9, the irreversible PM demagnetization just occurs. In Ref. [28], it is claimed that the maximum back-EMF can be achieved when the rotor slot opening ratio b_o/t is around 0.6. So, during the design process, the optimal rotor slot opening ratio can be directly applied without consideration of the PM demagnetization risk.

5. Geometric design of stator and rotor

5.1 Stator design

The geometrical parameters of stator and rotor are shown in **Figure 26**. The no-load flux of each winding pole could be calculated as:

$$\phi_m = 2\lambda_w l_{stk} B_m / \pi \tag{32}$$

where λ_w is the winding pitch. If full-pitch winding is adopted, the winding pitch is able to be written as:

$$\lambda_w = 2\pi r_g / 2P \tag{33}$$

Then, the no-load flux of each winding pole ϕ_m in Eq. (32) could change to:

$$\phi_m = 2r_g l_{stk} B_m / P \tag{34}$$

Defining the average flux density at the stator yoke as B_y, the stator yoke thickness h_y can therefore be deduced as:

$$h_y = \frac{\phi_m}{2B_y k_{stk} l_{stk}} = \frac{r_g B_m}{P B_y k_{stk}} \tag{35}$$

Similarly, defining the average flux density at the middle of stator tooth as B_t, the stator tooth width is able to be worked out:

$$w_t = \frac{\phi_m}{3SPP k_{stk} l_{stk} B_t} = \frac{4 r_g B_m}{Z_s k_{stk} B_t} \tag{36}$$

Moreover, in order to simultaneously maintain a relatively large torque density as well as reduce the risk of PM demagnetization, the PM thickness is recommended to be:

$$h_m = 4g \sim 6g \tag{37}$$

where g is the airgap length. Since the optimal torque density is often obtained when the slot opening ratio is approximate to 0.25 [28], the stator slot opening width w_o could be written as:

$$w_o = \pi (r_g + h_m) / 2Z_s \tag{38}$$

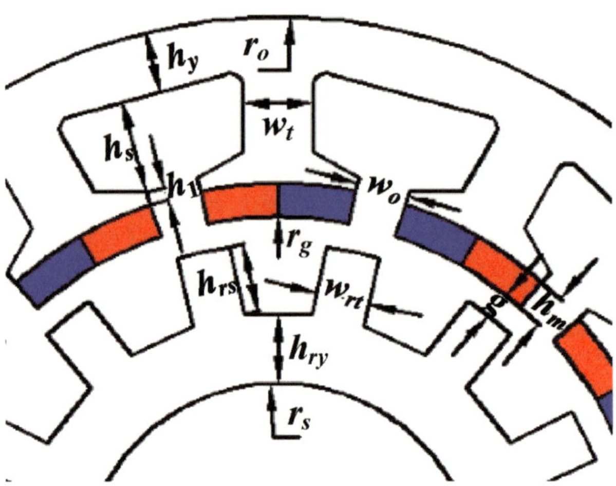

Figure 26.
Geometry of stator and rotor.

Then, next step is to calculate the stator outer radius r_o. Firstly, the total slot area of all the stator slots A_{slot} can be written based on the winding electric loading A_e and the current density J_e:

$$A_{slot} = 2\pi(r_g + h_m)A_e/J_e S_{fg} \tag{39}$$

where S_{fg} is the slot fill factor. Meanwhile, the total slot area of all the stator slots A_{slot} can be also derived out using the structural parameters:

$$A_{slot} = \pi(r_g + h_m + h_1 + h_s)^2 - \pi(r_g + h_m + h_1)^2 - Z_s w_t h_s \tag{40}$$

Combining the Eqs. (39) and (40), the slot depth h_s can be determined. Then, the stator outer radius r_o can be given as:

$$r_o = r_g + h_m + h_1 + h_s + h_y \tag{41}$$

5.2 Rotor design

Defining the average flux density of each rotor yoke and middle of rotor tooth as B_{ry}, and B_{rt}, respectively, the rotor yoke thickness h_{ry} and rotor tooth width w_{rt} are able to be achieved using the similar derivation procedure as Eq. (35) and Eq. (36). Finally, the h_{ry} and w_{rt} are given as:

$$h_{ry} = r_g B_m / Z_r B_{ry} k_{stk} \tag{42}$$

$$w_{rt} = 4 r_g B_m / Z_r k_{stk} B_{rt} \tag{43}$$

Then, the rotor slot depth h_{rs} is determined as:

$$h_{rs} = r_s + h_{ry} \tag{44}$$

6. Design methodology and evaluations

6.1 Design procedure

Based on the analytical equations and the investigations of key performances in the former parts, a quick and accurate analytical design of a FRPMM can be realized by following these procedures (as depicted in **Figure 27**):

1. Based on the performance investigations in **Figures 9–24**, the initial design values, including combination of stator slot and rotor slot number, electric loading, equivalent magnetic loading, airgap length, materials of active parts, etc. can be firstly selected.

2. Then, assuming an appropriate aspect ratio k_{lr}, the airgap radius r_g, and the stack length l_{stk} can be worked out using Eqs. (29) and (30).

3. Based on Eqs. (32–44), the detailed geometric parameters of the stator core and the rotor core are able to be obtained. Therefore, the stator outer diameter r_o and machine total length l_o can be finally determined.

4. After that, check if the stator outer diameter and the machine total length satisfy the required design specifications. If so, proceed to the FEA

verifications of machine performances. If not, reset the initial values such as combination of stator slot and rotor slot number, electric loading, equivalent magnetic loading, etc.

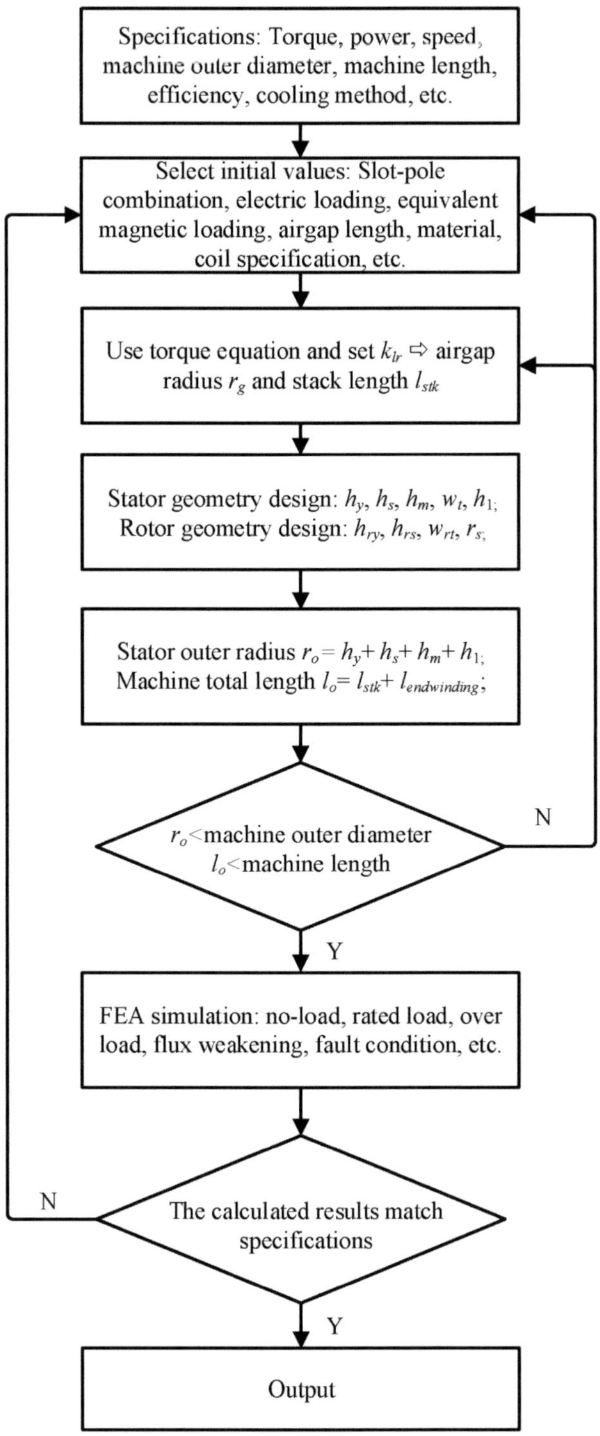

Figure 27.
Design flow of FRPMM.

5. Conducting FEA simulations, the electromagnetic performances such as back-EMF, average torque, pulsating torque, power factor, efficiency, etc. can be obtained. Check if all the performances satisfy the design specifications. If not, adjust the design parameters in the former steps and iterate the design flow until every output meets the requirement.

6. Finally, it is the result output.

6.2 Case study

In order show the effectiveness of the introduced analytical method, a FRPMM is designed based on the method. **Table 4** shows the specifications of the FRPMM, which mainly includes the rated torque, machine volume, cooling method, rated power, and speed. According to the rated torque, a design margin of 5% is suggested so as to make sure the torque output. Therefore, the requirement of the torque is 8.4 Nm for this design. Then, the combination of stator slots and rotor slots is determined in the first place. This combination is selected due to its high torque density and low pulsating torque, as shown in **Figure 13**. Then, since the cooling method is natural cooling, the electric loading and the equivalent magnetic loading are chosen as 300A/cm and 0.2 T, respectively. After that, based on the output torque value 8.4 Nm and Eq. (29), the airgap radius is determined as 38.5 mm. Furthermore, assuming the yoke flux density of stator core and rotor core as 1.0 T, and the teeth flux density of stator core and rotor core as 1.2 T, the detailed

Parameter	Value	Parameter	Value
Rated torque	8 Nm	Rated speed	300 rpm
Rotor inner diameter	32 mm	PM material	N38SH
Stator outer diameter	130 mm	Stack length	120 mm
Airgap length	0.6 mm	Iron material	50WW470
Cooling method	Natural cooling	Rated power	0.25 kW

Table 4.
Design specifications of a three-phase FRPMM.

	Parameter	Value	Parameter	Value
Stator	Outer diameter	124 mm	Inner diameter	79 mm
	Turns per phase	300	Teeth width	11.5 mm
	Slot number	12	Yoke thickness	6 mm
	Slot depth	13.5 mm	Yoke flux density	1.0 T
	Winding pole pair	1	Teeth flux density	1.1 T
Magnet	PM thickness	3 mm	Magnet width	7.8 mm
Rotor	Outer diameter	77.8 mm	Slot depth	10.4 mm
	Teeth flux density	1.2 T	Yoke thickness	12.5 mm
	Inner diameter	32 mm	Yoke flux density	1.0 T
	Teeth width	4 mm	Slot number	17

Table 5.
Design parameters of the FRPMM using the design method.

Parameter	Analytical design method	FEA
PM flux linkage	1.43 Wb	1.35 Wb
Back-EMF	44.8 V	42.4 V
Torque	8.4 Nm	7.97 Nm

Table 6.
Results comparison of the design method and 2D FEA.

Figure 28.
12-slot/17-pole FRPMM prototype: (a) stator; (b) rotor.

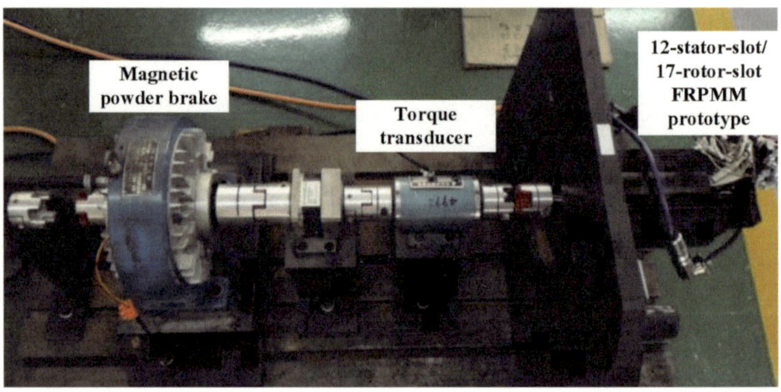

Figure 29.
Test bed of the FRPMM prototype.

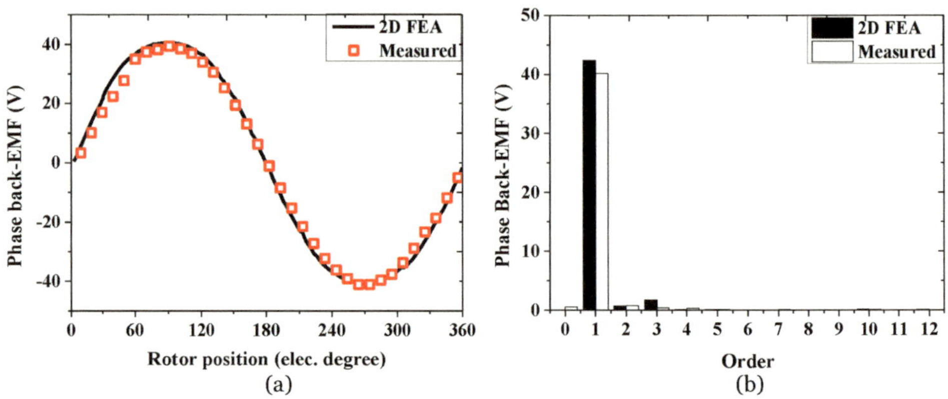

Figure 30.
Back-EMF waveforms at rated speed 300 rpm: (a) waveform; (b) FFT analysis.

geometric parameters can all be determined. At last, the stator outer diameter is worked out as 124 mm, which is less than the requirement 130 mm. So far, this design is effective. **Table 5** summarizes the design parameters of the FRPMM. Finally, in order to verify the accuracy of the proposed analytical design method, the FEA model is built, and the simulated performances are compared to the analytical designed values. It can be seen in **Table 6** that the FEA simulated results match well with the analytical method. More importantly, the simulated performance output satisfies the design specifications. Therefore, this analytical design is successful.

6.3 Experimental study

To verify the calculated results by the analytical method and FEA, the FRPMM prototype has been built. Its major parameters are listed in **Table 4**. The structure and test bed of the prototype are shown in **Figures 28** and **29**, respectively.

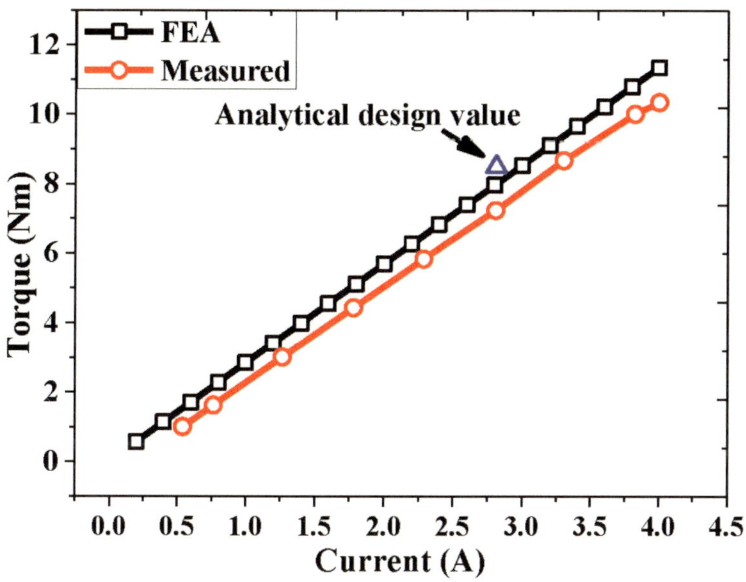

Figure 31.
Output torque vs. phase current.

Parameter	FEA	Experiment
Average torque at rated current	7.97 Nm	7.24 Nm
Torque per weight	0.66 Nm/kg	0.60 Nm/kg
Phase back-EMF magnitude at 300 rpm	42.4 V	41.2 V
THD of the phase back-EMF at 300 rpm	1.26%	2.63%
Total losses	99.5 W	116.7 W
Efficiency	60.3%	57.3%
Power factor	0.756	0.746

Table 7.
Result comparison of FEA and experiment of the FRPMM prototype.

Figure 30 compares the phase back-EMF waveform and spectrum at 300 rpm. It can be seen that the back-EMF waveforms are very sinusoidal. This is because the total harmonic distortion (THD) of FEA and experiments are only 1.26% and 2.63%, respectively. The sinusoidal back-EMF is inherent without any special design techniques such as skewing or pole shaping. Then, **Figure 31** shows the FEA simulated and experimental results of average torque at different winding current values. In addition, the analytical design value is also plotted as the blue triangle. It indicates that the simulated, analytical and experimental results have reached good agreements. Finally, **Table 7** compares the electromagnetic performances by FEA and experiments. Thus, the feasibility of the analytical design method can be seen.

7. Conclusions

The design of FRPMMs is usually based on time-stepping FEA, which are accurate but time-consuming. To save the design time meanwhile maintain the accuracy, this chapter proposes an analytical design method of FRPMMs. First, the sizing equation is derived, and then the dimensional parameters of stator and rotor are calculated. Finally, based on the above equations, an analytical design procedure is established. Moreover, in order to help to choose the initial design parameters in the sizing equation, including number of stator slots and rotor slots, airgap radius, electrical loading, and equivalent magnetic loading, their effects on the average torque, cogging torque, torque ripple, and power factor are investigated, providing reliable guidance for designers. At last, in order to make the introduced design methodology easier to understand, a FRPMM is designed and tested.

Acknowledgements

This work was supported by National Natural Science Foundation of China (NSFC) under Project Number 51807076, and Alexander von Humboldt Foundation.

Nomenclature

B_r	remanent flux density
μ_r	relative permeability of magnets
g'	effective airgap length considering PM thickness
h_m	PM height along the magnetization direction
SO	stator slot opening ratio (=slot opening width/slot pitch)
g	airgap length
r_g	airgap radius
N_s	number of series turns per phase
P	number of stator winding pole pairs
l_{stk}	active stack length
Z_r	number of rotor teeth
ω_m	mechanical angular speed of rotor
Z_s	number of stator teeth
SPP	slot per pole per phase
θ	angular position of rotor axis with respect to the axis of phase *a*
θ_s	particular position in the stator reference frame measured from the axis of phase *a*
PR	pole ratio (=rotor pole number/winding pole pair)

Author details

Yuting Gao[1]* and Yang Liu[2]

1 Karlsruhe Institute of Technology, Karlsruhe, Germany

2 Wuhan Institute of Marine Electric Propulsion, Wuhan, China

*Address all correspondence to: gyt626890@gmail.com

IntechOpen

© 2020 The Author(s). Licensee IntechOpen. This chapter is distributed under the terms of the Creative Commons Attribution License (http://creativecommons.org/licenses/by/3.0), which permits unrestricted use, distribution, and reproduction in any medium, provided the original work is properly cited.

References

[1] Upadhyay P, Sheth NK, Rajagopal KP. Effect of rotor pole arc variation on the performane of flux reversal motor. In: International Conference on Electrical Machines and Systems. 2007. pp. 906-911

[2] Sheth NK, Rajagopal KR. Performance of flux reversal motor at various rotor pole arcs. In: International Conference on Electrical Machines and Systems. 2007. pp. 1517-1522

[3] Kushwaha D, Dwivedi A, Reddy R, et al. Study of 8/12 flux reversal machine as an alternator. In: Eighteenth National Power Systems Conference. 2014. pp. 1-4

[4] Shin HK, Kim TH, Kim CJ. Demagnetization characteristic analysis of inset-type flux-reversal machines. In: 15th International Conference on Electrical Machines and Systems. 2012. pp. 1-4

[5] Vandana R, Fernandes BG. Mitigation of voltage regulation problem in flux reversal machine. In: Energy Conversion Congress and Exposition. 2011. pp. 1549-1554

[6] More DS, Fernandes BG. Novel three phase flux reversal machine with full pitch winding. In: International Conference on Power Electronics. 2007. pp. 1007-1012

[7] Ahn J, Choi JH, Kim S, et al. Parametric variance consideration in speed control of single-phase flux reversal machine. IET Electric Power Applications. 2008;**2**(4):266-274

[8] Kim TH, Won SH, Lee J. Finite element analysis of flux-reversal machine considering BEMF current of a switch-off phase and v-i characteristics of a transistor and a freewheeling diode. IEEE Transactions on Magnetics. 2006;**42**(4):1039-1042

[9] Wang C, Nasar SA, Boldea I. High speed control scheme of flux reversal machine. In: International Conference on Electric Machines and Drives. 1999. pp. 779-781

[10] Deodhar R, Andersson S, Boldea I, Miller T. The flux-reversal machine: A new brushless doubly-salient permanent-magnet machine. IEEE Transactions on Industry Applications. 1997;**33**(4):925-934

[11] Wang CX, Boldea I, Nasar SA. Characterization of three phase flux reversal machine as an automotive generator. IEEE Transactions on Energy Conversion. 2001;**16**(1):74-80

[12] Lee CHT, Chau KT, Liu C, et al. A new magnetless flux-reversal HTS machine for direct-drive application. IEEE Transactions on Applied Superconductivity. 2015;**25**(3):5203105

[13] Prakht V, Dmitrievskii V, Klimarev V, et al. High speed flux reversal motor for power tool. In: International Electric Drives Production Conference. 2016. pp. 306-311

[14] Bahrami H, Zabihi A, Joorabian M. A novel flux-reversal axial flux generator for high speed applications. International Conference on Power Electronics and Drives Systems. 2005;**2**: 1152-1155

[15] Prakht V, Prakht V, Sarapulov S, et al. A multipole single-phase SMC flux reversal motor for fans. In: International Conference on Electrical Machines. 2016. pp. 53-59

[16] Dmitrievskii V, Prakht V, Mikhalitsyn A. A new single-phase flux reversal motor with the cores made of

soft magnetic composite materials. In: International Conference on Electrical Machines and Systems. 2015. pp. 936-939

[17] Pellegrino G, Gerada C. Modeling of flux reversal machines for direct drive applications. In: European Conference on Power Electronics and Applications. 2011. pp. 1-10

[18] More DS, Kalluru H, Fernandes BG. Outer rotor flux reversal machine for rooftop wind generator. In: International Conference on IEEE Industry Application Society Annual Meeting. 2008. pp. 1-6

[19] Lee CHT, Chau KT, Liu C. Design and analysis of a cost-effective magnetless multiphase flux-reversal DC-field machine for wind power generation. IEEE Transactions on Energy Conversion. 2015;**30**(4): 1565-1573

[20] Li T, Sun Y, Li G, et al. Design of a novel double salient permanent magnet machine for wind power generation. In: International Conference on Electrical Machines and Systems. 2010. pp. 1053-1056

[21] Boldea I, Zhang L, Nasar SA. Theoretical characterization of flux reversal machine in low-speed servo drives-the pole-PM configuration. IEEE Transactions on Industry Applications. 2002;**38**(6):1549-1557

[22] Li W, Chau KT, Ching TW. A six-phase transverse-flux-reversal linear machine for low-speed reciprocating power generation. In: IEEE International Electric Machines and Drives Conference. 2015. pp. 618-623

[23] Gandhi A, Mohammadpour A, Sadeghi S. Doubled-sided FRLSM for long-stroke safety-critical applications. In: IEEE Industrial Electronics Society Annual Conference. 2011. pp. 4186-4191

[24] Chung S, Lee H, Sang M. A novel design of linear synchronous motor using FRM topology. IEEE Transactions on Magnetics. 2008;**44**(6):1514-1517

[25] Boldea I, Wang CX, Nasar SA. Design of a three-phase flux reversal machine. Electric Machines and Power Systems. 1999;**27**:849-863

[26] Kim TH, Lee J. A study of the design for the flux reversal machine. IEEE Transactions on Magnetics. July 2004; **40**(4):2053-2055

[27] Kim B, Lipo TA. Operation and design principles of a PM Vernier motor. IEEE Transactions on Industry Applications. March 2014;**50**(6): 3656-3663

[28] Gao Y, Qu R, Li D, et al. Design procedure of flux reversal permanent magnet machines. IEEE Transactions on Industry Applications. 2017;**53**(5): 4232-4241

[29] Zhu ZQ, Howe D. Instantaneous magnetic-field distribution in brushless permanent-magnet dc motors, Part III. Effect of stator slotting. IEEE Transactions on Magnetics. 1993;**29**(1): 143-151

[30] Gao Y, Qu R, Li D, et al. Torque performance analysis of three-phase flux reversal machines for electric vehicle propulsion. IEEE Transactions on Industry Applications. 2017;**53**(3): 2110-2119

Chapter 5

Predictive Direct Torque Control Strategy for Doubly Fed Induction Machine for Torque and Flux Ripple Minimization

Gopala Venu Madhav and Y.P. Obulesu

Abstract

The main drawback of Direct Torque Control (DTC) or Direct Power Control (DPC) is non-constant switching frequency; this drawback can be eliminated by employing predictive DTC. The predictive DTC technique is employed without much complicated online calculations by simply implementing constant switching times for active rotor voltage vectors to reduce torque and flux ripples and achieve constant switching frequency. The predictive DTC strategy has been implemented for RSC of Doubly Fed Induction Machine (DFIM). The performance of the proposed control methodology is compared with the classical DTC method under various operating conditions such as step change in torque, continuous variation of torque command, and the performance of DFIM near synchronous speed. It is found that the performance of the proposed predictive DTC strategy of DFIM is quite good compared to classical DTC strategy.

Keywords: predictive direct torque control, classical direct torque control, doubly fed induction machine, torque ripple, flux ripple

1. Introduction

In the field of renewable energy sources, wind energy is gaining much importance. The increase in the level of generation has two main restrictions, one is due to the limitations of the switching frequency of the power devices with respect to the power drive of the Doubly Fed Induction Machine (DFIM) and second one is the requirement of good dynamic torque performance, these restrictions are addressed by the proposed new Direct Torque Control (DTC) approach.

The foremost torque control methods proposed were classified into Field Oriented Control (FOC) techniques, Blaschke [1] and the direct control techniques. Then after, the concept of DTC, Takahashi and Ohmori [2] and Direct Self Control (DSC), Depenbrock [3] were introduced to achieve good steady-state and transient torque control conditions. Moreover, direct control techniques do not require current regulators, nor coordinate transformations or specific modulations like Pulse Width Modulation (PWM) or Space Vector Modulation (SVM) for pulse generation. The disadvantages are the lack of direct current control, torque control difficulties at very low speeds and especially variable switching frequency behavior.

The last important drawback put forward several authors in recent years, thus, methods like Direct Mean Torque Control (DMTC), [4] and Direct Torque Control based on Discrete Space Vector Modulation (DTC–DSVM), [5], have already achieved constant switching frequency. Furthermore, the methods proposed in Kang and Sul [6, 7] have extended the solution to reduce torque ripple at the same time imposing the switching frequency, as well as for different voltage-source multilevel topologies.

Further the concept of DTC have been applied to the brushless doubly fed induction machine, [8], or the equivalent Direct Power Control (DPC) strategy for several grid connected converter applications, [9].

In this paper, the analysis on the Doubly Fed Induction Machine (DFIM), which is a common solution for variable speed wind turbines, is discussed. The control methods like FOC have been performed by many authors, for example [10]. DTC and DPC methods without switching frequency imposition have also been carried out in Gomez & Amenedo [11], Datta & Ranganathan [12], while the DPC at constant switching frequency has also been developed in [13].

These mentioned direct control techniques that achieve constant switching frequency behavior are based on predictive control with a prediction horizon equal to one sample period. In this paper, predictive DTC technique for the DFIM will be employed based on [13] and the performance is compared with the DTC strategy, at constant switching frequency and with reduced torque and flux ripples criteria. This control technique is based on a prediction of the torque and the flux evolution of the DFIM. Hence, the new Predictive DTC (PDTC) strategy presented in this paper is based on a direct control of the electromagnetic torque and the rotor flux of the machine. Simulation results are presented and discussed, and at last, the results show that the predictive DTC technique presents good dynamic response compared to classical DTC concept.

In this Chapter, predictive DTC technique for the DFIM has been implemented and its performance is compared with the classical DTC of DFIM based on certain parameters like constant switching frequency, torque ripple and flux ripple. This control technique is based on a prediction of the torque and the flux evolution of the DFIM. To validate the proposed control scheme, results are presented. From these results, it is observed that the predictive DTC technique gives good dynamic response compared to classical DTC concept.

The control strategy even reduces the switching losses of the converter and reduces the electromagnetic torque and flux ripples at low switching frequency even under variable speed operating conditions.

In Section 1, the introduction of the Chapter is given.

In Section 2, Contributions and Novelty of this Chapter is explained.

In Section 3, Modeling of the DFIM is given.

In Section 4, the basic control principle of predictive DTC is explained and also implementation of the proposed predictive DTC strategy of DFIM along with selection of rotor voltage vectors for constant switching frequency and reduction of switching power losses are described.

In Section 5, Results are presented to validate the proposed control strategy.

In Section 6, the Conclusions of the Chapter are described.

In Section 7, the summary of the Chapter is given.

2. Contributions and novelty

The main contributions of this Chapter are as follows:

- Implementation of a new predictive Direct Torque Control (DTC) strategy of the Doubly Fed Induction Machine (DFIM) is presented which is designed to operate at a low constant switching frequency.

- The proposed DTC method effectively reduces the torque and flux ripples at low switching frequency, even under variable speed operation conditions.

- Results are presented to validate the proposed control strategy.

The novelty of this Chapter is that the predictive DTC have been developed to control the different parameters of DFIM, to improve its performance during transient, steady state, tracking behavior and operation near synchronous speed. All these cases are thoroughly investigated.

3. Modeling of DFIM

The DFIG model adopted is the qd0 rotating reference frame. It is because the model of DFIM is quite suitable with this frame of reference during transients. The transient solution of the DFIM model is possible because of the transformation from abc to qd0 by which the differential equations with time-varying inductances is converted into differential equations with constant inductances,

$$\begin{aligned} v_{qs} &= R_s i_{qs} + \omega_s \psi_{ds} + \frac{d\psi_{qs}}{dt} \\ v_{ds} &= R_s i_{ds} - \omega_s \psi_{qs} + \frac{d\psi_{ds}}{dt} \end{aligned} \tag{1}$$

(or) simply (1) can be written as

$$\bar{v}_s^s = R_s \bar{i}_s^s + \frac{d\bar{\psi}_s^s}{dt} \tag{2}$$

Similarly, the q and d-axis rotor voltages referred to the stator are given by,

$$\begin{aligned} v'_{qr} &= R'_r i'_{qr} + (\omega_s - \omega_r)\psi'_{dr} + \frac{d\psi'_{qr}}{dt} \\ v'_{dr} &= R'_r i'_{dr} - (\omega_s - \omega_r)\psi'_{qr} + \frac{d\psi'_{dr}}{dt} \end{aligned} \tag{3}$$

(or) simply (3) can be written as

$$\bar{v}_r^s = R_r \bar{i}_r^s + \frac{d\bar{\psi}_r^s}{dt} - j\omega_m \bar{\psi}_r^s \tag{4}$$

The stator and rotor fluxes can be calculated by using (5) and (6) can be used to calculate the magnitudes.

$$\overline{\psi}_s = L_s\overline{I}_s + L_m\overline{I}_r$$
$$\overline{\psi}_r = L_m\overline{I}_s + L_r\overline{I}_r \tag{5}$$

$$|\psi_s| = \sqrt{\psi_{ds}^2 + \psi_{qs}^2}$$
$$|\psi_r| = \sqrt{\psi_{dr}^2 + \psi_{qr}^2} \tag{6}$$

The electromagnetic torque of DFIM is given in (7).

$$T_{em} = \frac{3}{2}p\mathrm{Im}\{\overline{\psi}_s^{\wedge}.\overline{i}_s\}\ (\mathrm{or})\ \frac{3}{2}p\left(\psi_{s\alpha}i_{s\beta} - \psi_{s\beta}i_{s\alpha}\right)$$
$$(\mathrm{or})\ \frac{3}{2}p\left(\psi_{ds}i_{qs} - \psi_{qs}i_{ds}\right)\ (\mathrm{or}) \approx \frac{3}{2}\frac{p}{2}L_m(i_{dr}'i_{qs} - i_{qr}'i_{ds}) \tag{7}$$

The active and reactive powers are given by:

$$P_s = \frac{3}{2}\mathrm{Re}\{\overline{v}.\vec{i}^{\wedge}\}\ (\mathrm{or})\ \frac{3}{2}(v_\alpha i_\alpha + v_\beta i_\beta)\ (\mathrm{or})\ \frac{3}{2}(v_d i_d + v_q i_q)\overline{\psi}_r = L_m\overline{I}_s + L_r\overline{I}_r$$

$$Q_s = \frac{3}{2}\mathrm{Im}\{\vec{v}.\vec{i}^{\wedge}\}(\mathrm{or})\frac{3}{2}(v_\beta i_\alpha - v_\alpha i_\beta)(\mathrm{or})\frac{3}{2}(v_q i_d - v_d i_q) \tag{8}$$

4. Description of proposed predictive direct torque control strategy for torque and flux ripple minimization

The main drawback of classical DTC scheme is nonconstant switching behavior; it is avoided by the proposed predictive Direct Torque Control strategy. The constant switching behavior is achieved by increasing slightly the complexity of control strategy.

A sequence of three voltage vectors will be introduced at a constant switching period of which two are active vectors always followed by a zero vector, in order to reduce the ripples of both directly controlled variables compared to selection of four rotor voltage vectors depending on the position of rotor flux in classical DTC scheme. For that purpose, ripple reduction criteria based on a prediction of the electromagnetic torque and rotor flux evolution over time is implemented which is derived from (7) and (10) [13] and illustrated in **Figure 1(a)**.

The basic principles of the predictive DTC scheme are firstly, constant switching period 't_s' is defined. In predictive DTC, the control procedure is discretized unlike in DTC scheme, which is based on time domain. Secondly, at steady state condition, by taking the electromagnetic torque and flux amplitude errors to be minimized, the three different rotor voltage vectors are injected during t_s. Note that according to the chosen three vector sequence together with the specified time intervals for each vector, the electromagnetic torque and flux evolutions within the switching period can be different. Thirdly, this procedure is repeated at constant period t_s.

The Doubly Fed Induction Machine is modeled using (2), (4)–(6). The torque is calculated by using (7) in terms of complex conjugate of rotor flux and stator flux.

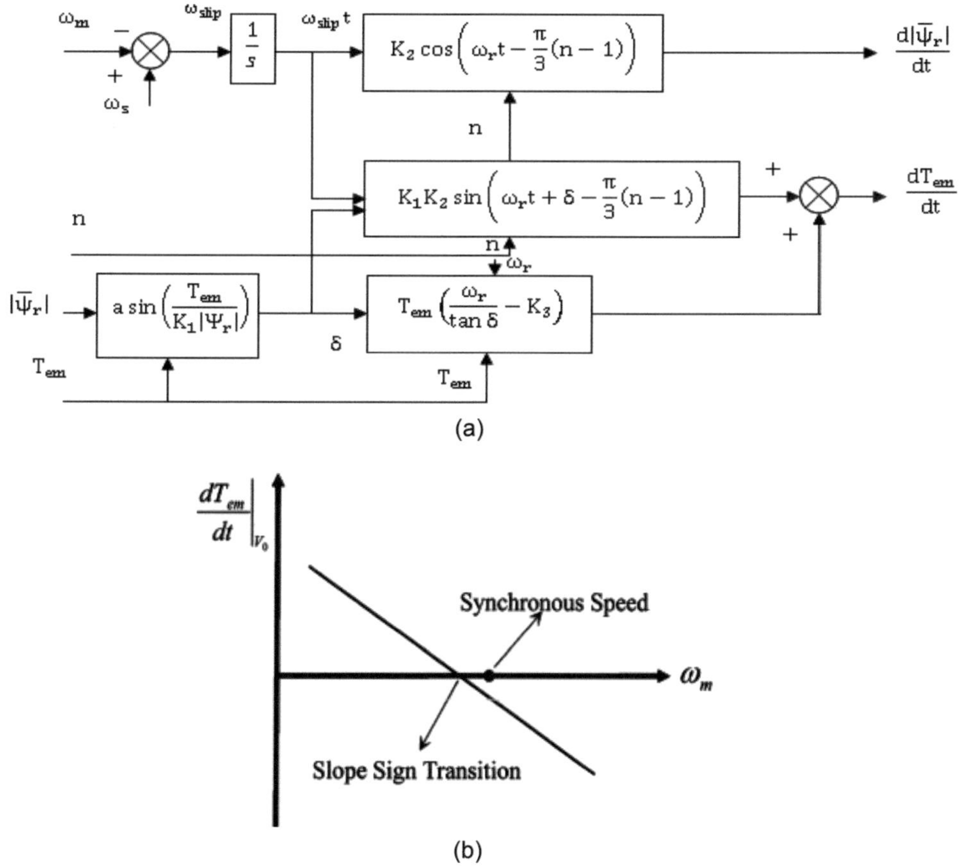

Figure 1.
(a) Simplified prediction scheme of torque and rotor flux of DFIM. (b) Variation of electromagnetic torque with zero vectors at constant torque and rotor flux.

4.1 Effect of voltage vector on the DFIM

Using (2), (4)–(7), the predictive expressions for torque and rotor flux are given by [2, 13]

$$\frac{d|\bar{\psi}_r|}{dt} = \frac{1}{|\bar{\psi}_r|}\left[\left(\frac{R_r L_m}{\sigma L_s L_r}\right)\mathrm{Re}\{\bar{\psi}_r.\bar{\psi}_s\} - \left(\frac{R_r}{\sigma L_r}\right)|\bar{\psi}_r|^2 + \mathrm{Re}\{\bar{\psi}_s.\bar{v}_r\}\right] \tag{9}$$

$$\frac{dT_{em}}{dt} = \frac{3}{2}p\frac{L_m}{\sigma L_s L_r}\left[\left(\frac{R_s}{\sigma L_s} + \frac{R_r}{\sigma L_r}\right)\mathrm{Im}\{\bar{\psi}_r.\hat{\bar{\psi}}_s\} - \omega_m \mathrm{Re}\{\bar{\psi}_r.\bar{\psi}_s\} + \mathrm{Im}\{\bar{v}_s.\hat{\bar{\psi}}_r\} + \mathrm{Im}\{\bar{\psi}_s.\hat{\bar{v}}_r\}\right] \tag{10}$$

The below space vector representations are used in order to analyze the expressions (9) and (10).

$$\bar{\psi}_r = |\bar{\psi}_r| e^{j\omega_s t} \tag{11}$$

$$\bar{\psi}_s = |\bar{\psi}_s| e^{j(\omega_s t + \delta)} \tag{12}$$

$$\bar{v}_s = |\bar{v}_s| e^{j\left(\omega_s t + \left(\frac{\pi}{2}\right) + \delta\right)} \tag{13}$$

$$\bar{v}_r = \frac{2}{3} V_{DC} e^{j(\omega_m t + \pi/3(n-1))} \tag{14}$$

where *n* is representation of sectors from 0 to 7. In (9), if Eqs. (11)–(14) are substituted, then

$$\frac{d|\bar{\psi}_r|}{dt} = \left[\left(\frac{R_r L_m}{\sigma L_s L_r}\right)|\bar{\psi}_s|\cos\delta - \left(\frac{R_r}{\sigma L_r}\right)|\bar{\psi}_r| + \frac{2}{3}V_{DC}\cos\left(\omega_{slip}t - \frac{\pi}{3}(n-1)\right)\right] \tag{15}$$

where

$$\omega_{slip} = \omega_s - \omega_m \tag{16}$$

Eq. (15) comprises of a cosine term and two constant terms, the cosine term with depends on DC bus voltage and it indicates that the cosine term of rotor flux variation is constant for zero vectors and only depends on active vectors.

Eq. (10) can be simplified considering the stator flux vector module which is nearly constant.

$$\frac{dT_{em}}{dt} = T_{em}\left(\frac{\omega_{slip}}{\tan\delta} - \left(\frac{R_s}{\sigma L_s} + \frac{R_r}{\sigma L_r}\right)\right) + p\frac{L_m}{\sigma L_s L_r}V_{DC}|\bar{\psi}_s|\sin\left(\omega_{slip}t + \delta - \frac{\pi}{3}(n-1)\right) \tag{17}$$

Eq. (17) is similar to rotor flux derivation with only one constant and one sine term.

Equations (15) and (17) are used practically as shown in **Figure 1(a)**, instead of (9) and (10). As shown in **Figure 1(a)**, the torque and flux derivatives depends on only four terms when considering the DC bus voltage, stator voltage and flux as constant magnitudes. In the expression (17), the constant term depends on T_{em} and, that means it depends on machine operating condition, which can be positive or negative as shown in **Figure 1(b)**. Because of this reason only, the phase shift order of each active vector varies, whereas, it is not like that for rotor flux derivative. The key point is knowing the slopes or the derivatives of torque and flux for each of the rotor voltage vector based on **Figure 1(a)** and **Table 1** and considering these slopes to be constants within the specific time in the given switching period, t_s, the torque and flux ripples can be maintained within the limits.

With constant values

$$K_1 = \frac{3}{2}p\frac{L_m}{\sigma L_s L_r}|\bar{\psi}_s| \tag{18}$$

$$K_2 = \frac{2}{3}V_{DC} \tag{19}$$

$$K_3 = \frac{R_s}{\sigma L_s} + \frac{R_r}{\sigma L_r} \tag{20}$$

From the **Figure 1(b)**, it can be seen that the slope of torque derivative varies proportionally to the slip speed provided at fixed torque and rotor flux operating conditions and also near synchronous speed, the slope of the zero vector becomes smaller obeying the fact that from the expression (17) the zero vector produce small torque variation. This fact implies that near synchronous speed the amplitude of rotor voltage vector is small demanding the condition of zero vectors. As shown in **Figure 1(b)**, this transition from positive to negative torque slope is different to the synchronous speed.

4.2 Selection of first rotor voltage vector

The first vector is selected depending on the errors of torque and flux and the sector where the rotor flux lies, the look up table for vector selection is shown in **Table 1**. From the **Table 1**, it is noticed that the required rotor voltage vector should produce either positive or negative slope variation depending on the output of the torque or flux hysteresis comparators that is either 1 or −1, respectively.

4.3 Selection of second and third vector rotor voltage vector

The first vector is actually selected based on the classical DTC; the second vector is selected such that it is always followed by zero vectors in order to reduce the torque and flux ripples based on the predictive DTC strategy.

From **Figure 2**, it can be seen that from the derivative calculations block the required values of torque and flux evolutions depending on the each rotor voltage selected are fed to the ripple reduction criteria block, in which based on the slopes calculations by (15) and (17), the required rotor voltage vectors which are selected are active for the time period of this constant slopes. This constant time period of the constant slopes is fed to the switching table, where in the switching operation of rotor voltages are chosen in such a way that it reduces the switching losses.

From (15) and (17), it can be inferred that the rotor flux is constant for zero vector and it produces opposite sign for torque variation for first two active vectors. These two active vectors along with zero vectors are useful to control torque and rotor flux. By the two active vectors the rotor flux as one vector produce positive slope, the other vectors produce negative slope because of this the flux ripple is not eliminated completely as compared to torque ripple and further it affects the rotor and stator currents. **Table 2** shows the selection of second active vector, after the first active vector is selected. It clearly shows that, one vector cannot be selected, as the flux would have a very big or small variation, which leads to poor quality of flux output.

4.4 Reduction of switching power losses

The right choice of zero vectors V_0 and V_7 implies that there is reduction in switching power loss of the converter. Two different switching sequences exist for each pair of required active vectors, which allows the commutations of the converter to be reduced. **Table 3** shows the correct sequence of vectors, which allows only four commutations per switching period t_s. The candidate sequences are such

		Error of electromagnetic torque	
		1	**−1**
Error value of rotor flux	1	$V_{(n-1)}$	$V_{(n+1)}$
	−1	$V_{(n-2)}$	$V_{(n+2)}$

n = sector.

Table 1.
Selection of voltage vectors [13].

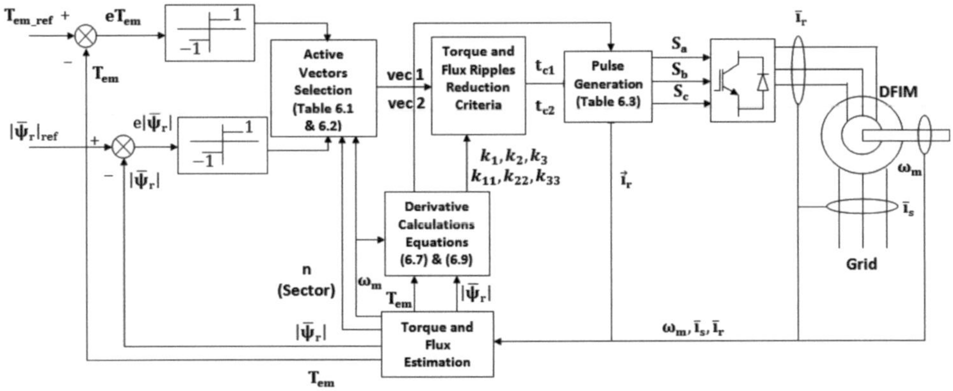

Figure 2.
Block diagram of proposed predictive direct torque control strategy of DFIM.

	Active vectors
Below synchronous speed	$V_{(n+1)}, V_{(n+2)}$
Above synchronous speed	$V_{(n-1)}, V_{(n-2)}$

n = sector.

Table 2.
Selection of active vectors under steady-state [13].

Active vectors	Sector	Switching sequences	Zero vector
V_1-V_2	6(subs)	100-110-111	V_7
	3(hypers)	110-100-000	V_0
V_2-V_3	1(subs)	110-010-000	V_0
	4(hypers)	010-110-111	V_7
V_3-V_4	2(subs)	010-011-111	V_7
	5(hypers)	011-010-000	V_0
V_4-V_5	3(subs)	011-001-000	V_0
	6(hypers)	001-011-111	V_7
V_5-V_6	4(subs)	001-101-111	V_7
	1(hypers)	101-001-000	V_0
V_6-V_1	5(subs)	101-100-000	V_0
	2(hypers)	100-101-111	V_7

subs – sub synchronous speed operation, hypers – super synchronous speed operation.

Table 3.
Sequence of vector selection for reduction of switching power losses under steady-state [13].

that it reduces the switching power losses by transmitting the smallest current values, which is the main task of the proposed control strategy.

5. Results and discussion

The predictive DTC control scheme of DFIM has been implemented using MATLAB/Simulink. The specifications of DFIM are given in **Table 4**. The

Doubly fed induction machine	
Specifications	**Ratings**
Rated power	2.6 MW
Rated voltage	690 V
Synchronous speed	1500 rpm
Frequency	50 Hz
Number of pairs of poles	2
Stator to rotor turns ratio	0.34
Base voltage	398.4 V
Base current	2175 A
Base angular frequency	157 rads^{-1}
Base power	2.6 MW
Stator resistance	0.0108 pu
Stator leakage inductance	0.102 pu
Magnetizing inductance	3.362 pu
Rotor resistance referred to stator	0.0121 pu
Rotor leakage inductance referred to stator	0.11 pu
Inertia constant(H)	0.5 s
Friction coefficient(F)	0.05479 pu
DC-link capacitance	5e–3F
DC-link voltage	1200 V
Wind turbine	
Number of blades	3
Rotor diameter	70 m
Hub height	84.3 m
Turbine total Inertia	4.4532 e5Kgm^2
Stiffness constant (B)	2 pu
Mutual damping	1.5 pu
Cut-in wind speed	3 ms^{-1}
Cut-out wind speed	25 ms^{-1}
Rated wind speed	15 ms^{-1}

Table 4.
Specifications of DFIM and wind turbine.

conditions of steady state, transient, tracking behavior, and performance near synchronism of DFIM are examined, which are given in three subsections.

5.1 Transient and steady state analysis of DFIM

The performance of proposed control strategy of DFIM is analyzed for steady state and transient conditions. For the transient conditions, the step change in electromagnetic torque i.e., from −0.4 pu to +0.4 pu at 0.6 s is considered. That means, from generator mode (negative torque) to motoring mode (positive torque), with constant switching frequency of 1 kHz, with speed reference of 1350 rev/min, and DC-link voltage of 1200 V. The DFIM is under steady state operation up to 0.6 s with torque of −0.4 pu and at 0.6 s the DFIM enters into transient state, and again it reaches its steady state value of 0.4 pu.

The response of stator currents in stationary reference frame are shown in **Figure 3(a)**. Therefore, there are two waveforms which refer to α, β components of stator currents. From the **Figure 3(a)**, it is observed that there are no over currents in stator, which indicates the effectiveness of the proposed control scheme even at sudden variation in torque demand. This is possible because of selection of proper rotor voltage vectors with their respective time intervals.

Figure 3(b) shows the response of developed torque for the proposed strategy and classical DTC strategy (not expressed in p.u. value), from the figure, it is noticed that the torque response of the DFIM closely followed the torque command and also torque ripple is zero.

From the **Figure 3(c)**, it can be seen that stator active power has good dynamic response when the reference torque is changed suddenly. From the figure, it is observed that the stator active power follows the torque demand to make the DFIM to develop the torque to match its reference value.

The response of the stator flux is shown in **Figure 3(d)**, from the figure, it is clearly noticed that the stator flux response remains constant which is not affected by variation in torque command.

The response of the rotor flux is shown in **Figure 3(e)**, from the figure, it is observed that the rotor flux response is also not affected by change in reference torque and also the rotor flux response is sinusoidal in nature which is not distorted due to sudden change in torque command.

Figure 3(f) shows response of the rotor speed of DFIM. From the **Figure 3(f)**, it is observed that there is decrease in rotor speed due to step change in reference torque but decrease in rotor speed is very small.

The response of rotor currents is shown in **Figure 3(g)**. From the **Figure 3(g)**, it is observed that there are no over currents in rotor, which indicates the effectiveness of the proposed control scheme even at sudden variation in torque demand. This is possible because of selection of proper rotor voltage vectors with their respective time intervals.

5.2 Performance analysis of DFIM during variable torque behavior

In this section, the proposed control scheme of DFIM is investigated for variable speed operation implying wind energy applications, and at the same instant, the torque reference may also vary respectively with speed of DFIM. This kind of behavior of wind energy system is called as tracking behavior. At this condition, the actual values of the DFIM should follow the reference values as closely as possible and this is clearly guaranteed by proposed control strategy which can be seen clearly through the results presented in this section. To explore the tracking behavior of DFIM, sinusoidally varying reference torque with amplitude of 0.4 pu and frequency of 3 Hz is set to the DFIM. By this set reference torque, the DFIM operates in generating and motoring modes. In this mode of operation, the other parameters of DFIM are same as mentioned in Section 5.1.

The predictive DTC strategy has good tracking behavior and it is confirmed that the reduction in torque and flux ripples is achieved as there are absolute absence of over currents and reduced ripples in stator currents as shown in **Figure 4(a)**. From the **Figure 4(a)**, it is noticed that there is continuous increase and decrease in the amplitude of stator currents for maintaining the consistency due to variable behavior of torque command. The stationary reference frame stator currents are clearly noticed in **Figure 4(a)**.

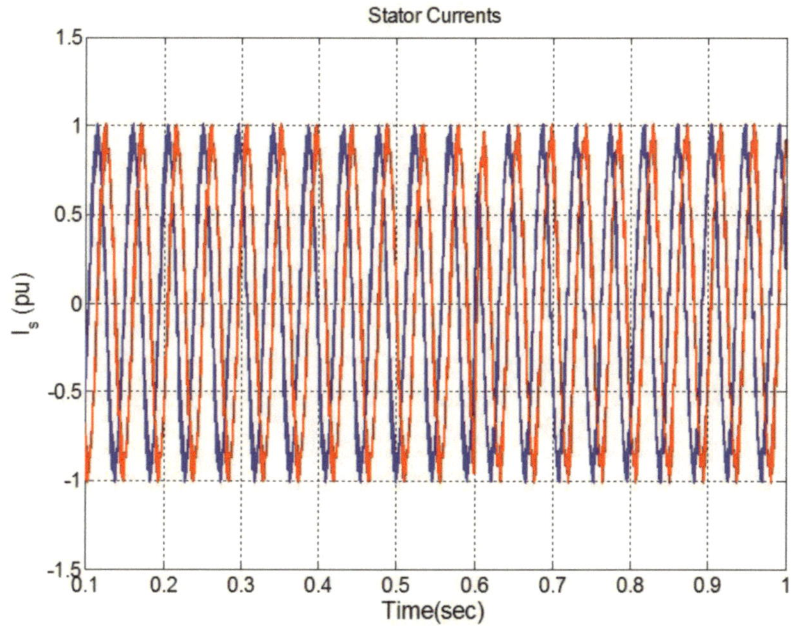
Stator Currents
I_s (pu)
Time(sec)

(a)

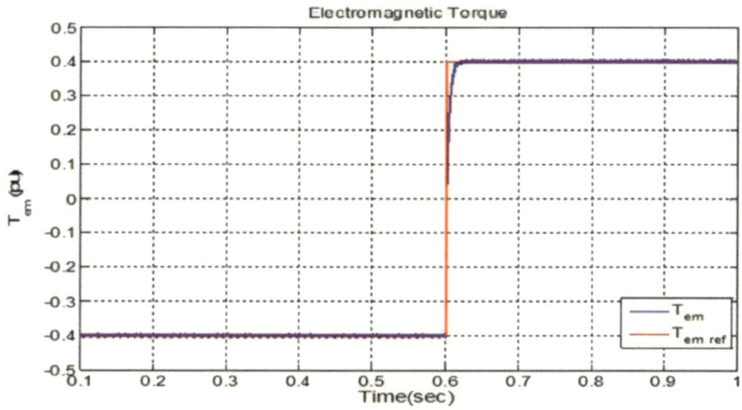
Electromagnetic Torque
T_{em} (pu)
Time(sec)
T_{em}
$T_{em\ ref}$

(i)

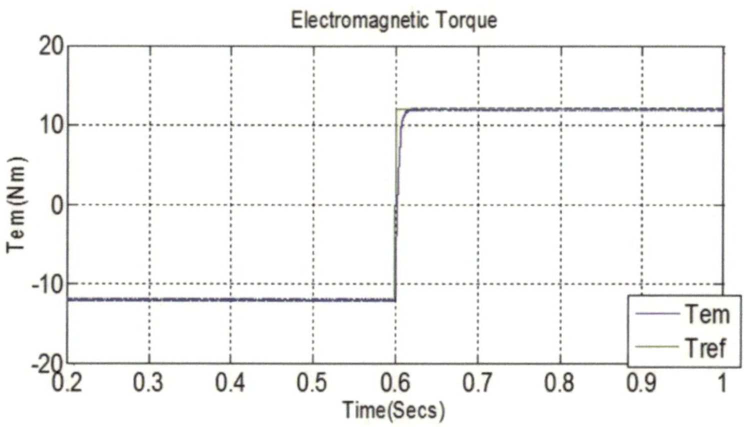
Electromagnetic Torque
Tem(Nm)
Time(Secs)
Tem
Tref

(ii)

(b)

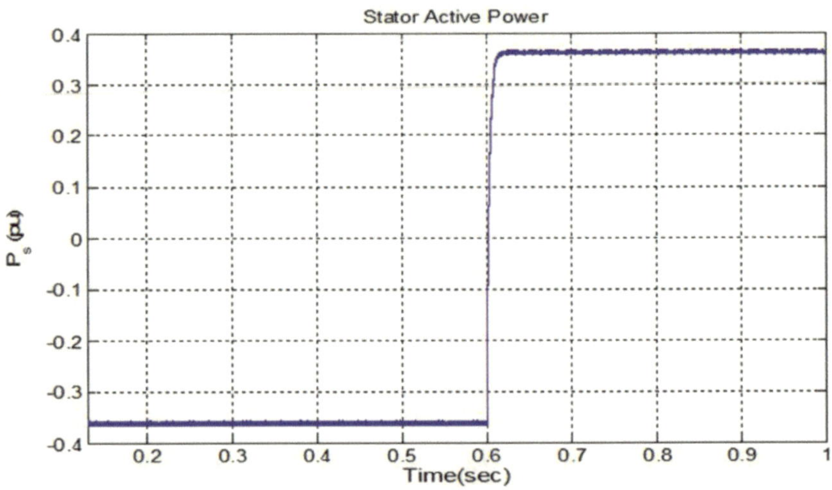
Stator Active Power
P_s (pu)
Time(sec)

(c)

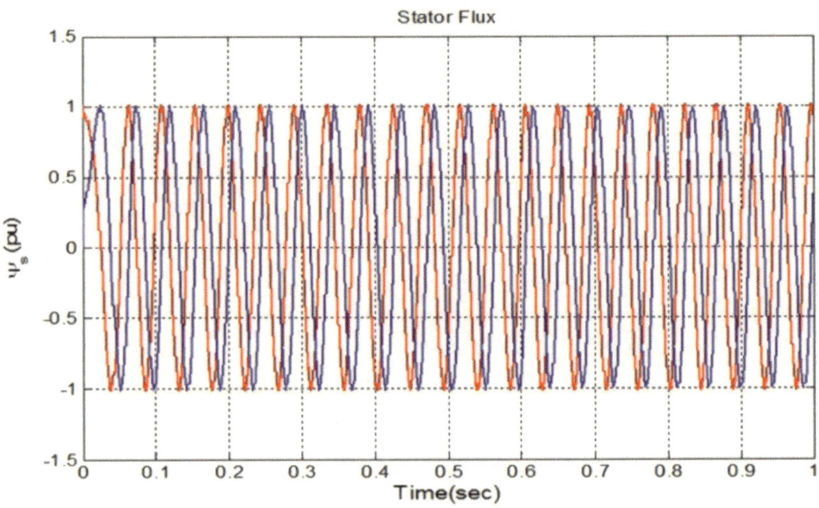
Stator Flux
ψ_s (pu)
Time(sec)

(d)

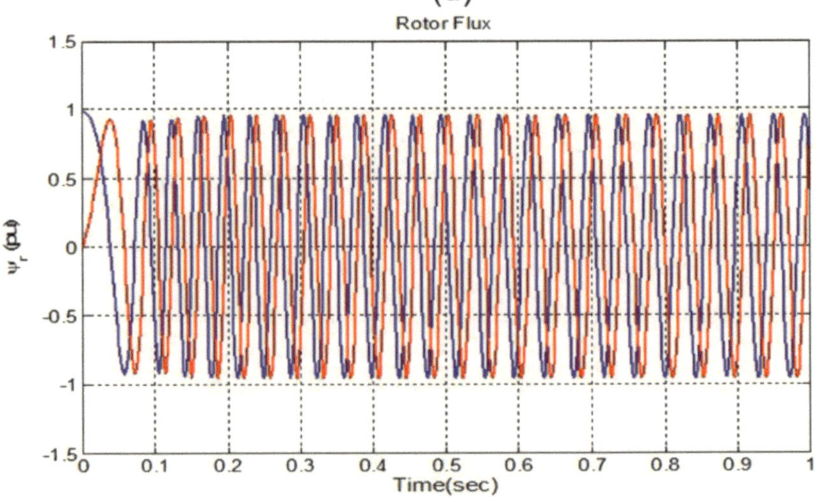
Rotor Flux
ψ_r (pu)
Time(sec)

(e)

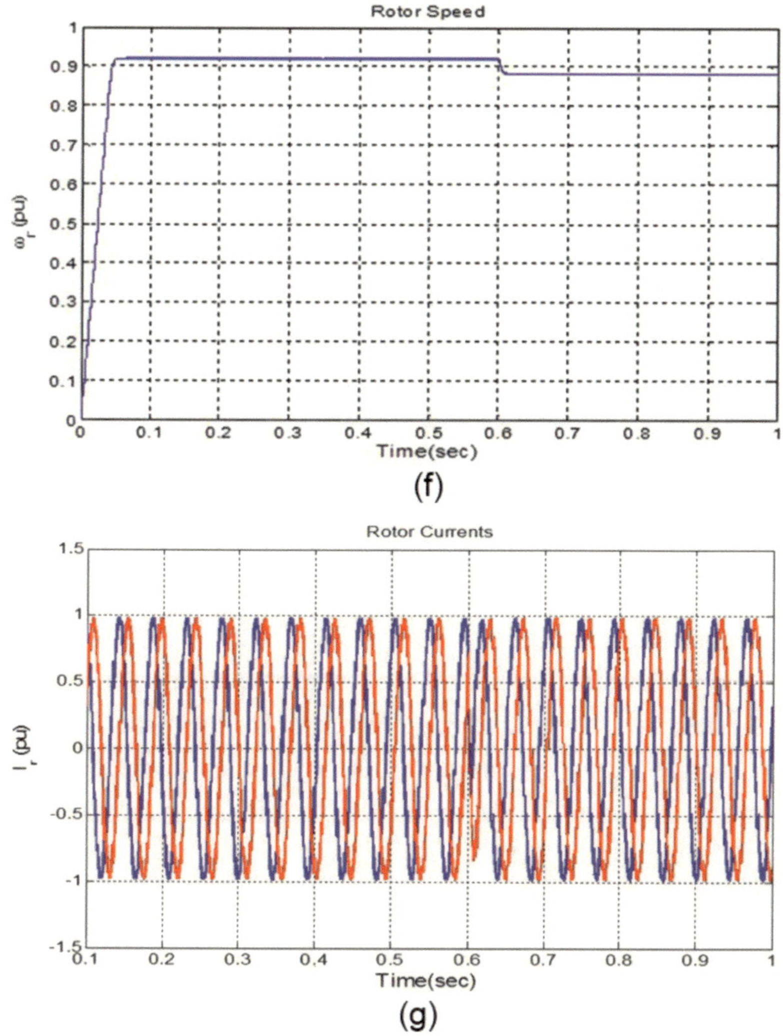

Figure 3.
(a) Response of stator currents of DFIM for step change in T_{em} from −0.4 pu to 0.4 pu at 0.6 s. (b) Torque response of DFIM for step change in T_{em} from −0.4 pu to 0.4 pu at 0.6 s, (i) proposed strategy (ii) classical DTC. (c) stator active power of DFIM for step change in T_{em} from −0.4 pu to 0.4 pu at 0.6 s. (d) Stator flux response of DFIM for step change in T_{em} from −0.4 pu to 0.4 pu at 0.6 s. (e) Response of rotor flux of DFIM for step change in T_{em} from −0.4 pu to 0.4 pu at 0.6 s. (f) Rotor speed response of DFIM for step change in T_{em} from −0.4 pu to 0.4 pu at 0.6 s. (g) Response of rotor currents of DFIM for step change in T_{em} from −0.4 pu to 0.4 pu at 0.6 s.

The torque produced by the DFIM follows as closely as the reference torque, which indicates good tracking behavior of the proposed control scheme comparative to classical DTC (not expressed in p.u. value), it can be seen in **Figure 4(b)**.

Stator flux and rotor flux responses of DFIM are shown in **Figure 4(c)** and **(d)** respectively. From the **Figure 4(c)** and **(d)**, it is observed that the variation in torque command is not having any influence on the stator and rotor fluxes of DFIM.

Figure 4(e) shows the rotor speed response of DFIM. From the **Figure 4(e)**, it is observed that there is continuous variation in rotor speed; of course, this variation is

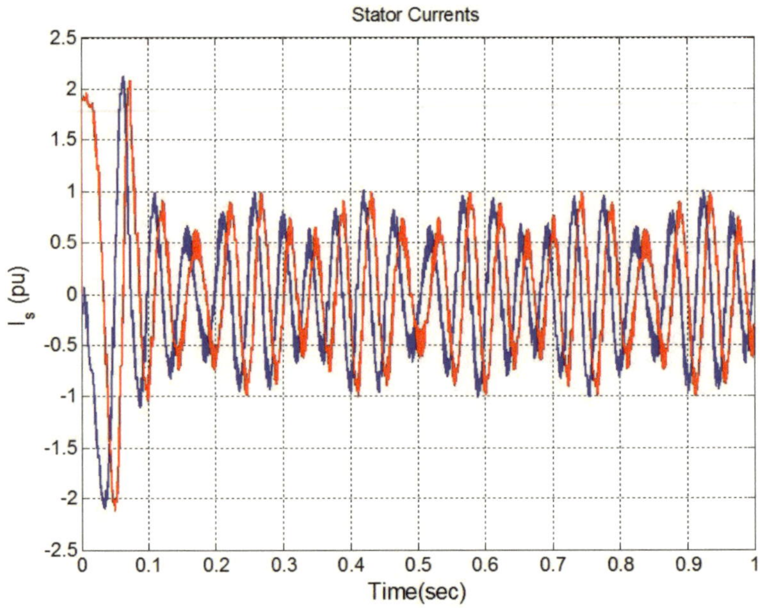
Stator Currents
I_s (pu)
Time(sec)
2.5
2
1.5
1
0.5
0
-0.5
-1
-1.5
-2
-2.5
0
0.1
0.2
0.3
0.4
0.5
0.6
0.7
0.8
0.9
1

(a)

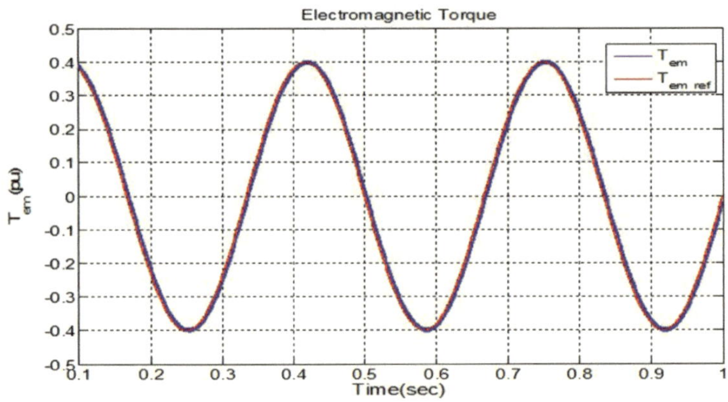
Electromagnetic Torque
T_em
T_em ref
T_em (pu)
Time(sec)

(i)

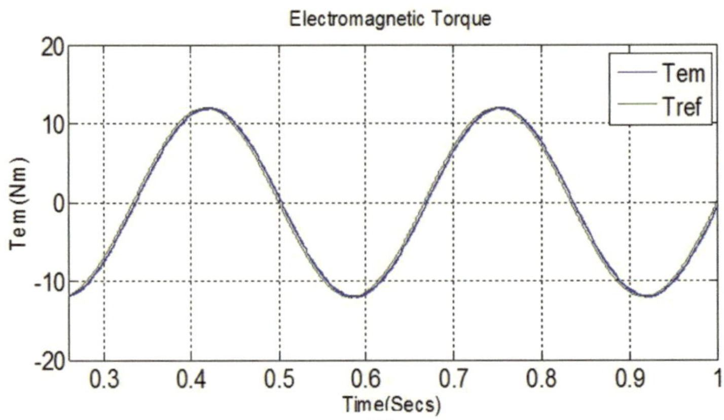
Electromagnetic Torque
Tem
Tref
Tem (Nm)
Time(Secs)

(ii)

(b)

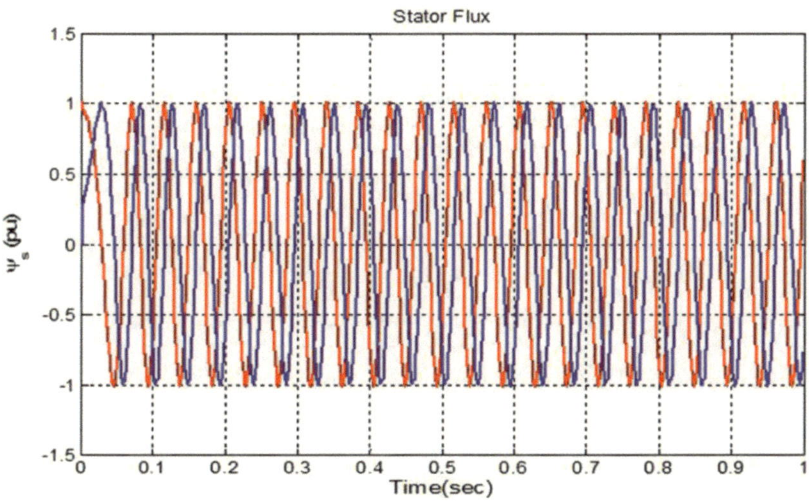
Stator Flux
ψs (pu)
Time(sec)

(c)

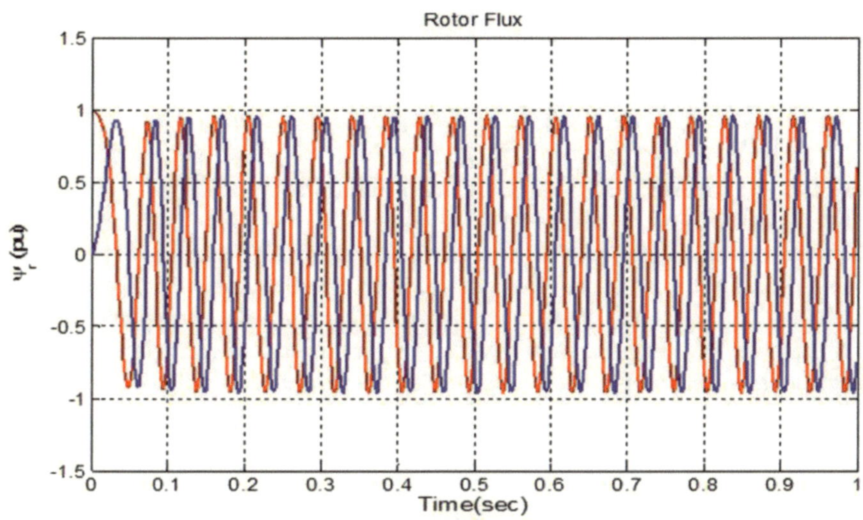
Rotor Flux
ψr (pu)
Time(sec)

(d)

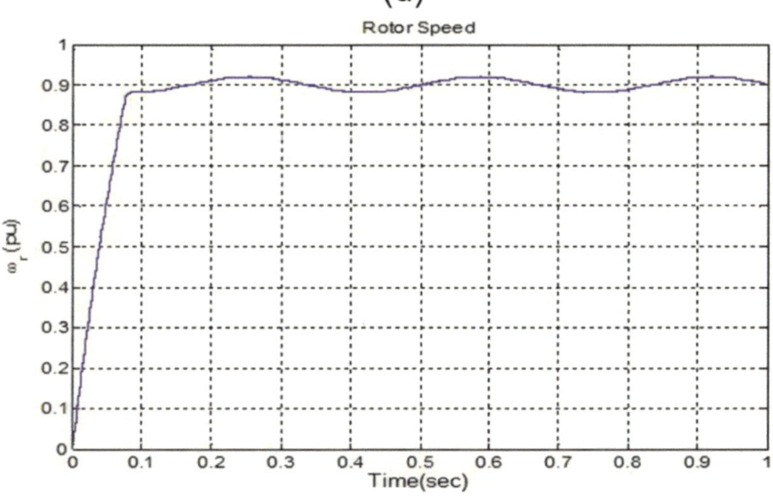
Rotor Speed
ωr (pu)
Time(sec)

(e)

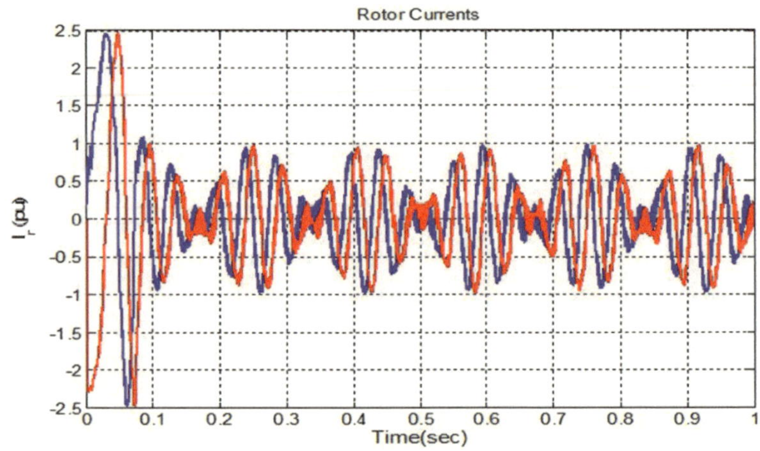

(f)

Figure 4.
(a) Response of stator currents of DFIM with variable torque command. (b) Response of developed torque of DFIM with variable torque command (i) proposed strategy (ii) classical DTC. (c) Response of stator flux of DFIM with variable torque command. (d) Response of rotor flux of DFIM with variable torque command. (e) Response of rotor speed of DFIM with variable torque command. (f) Response of rotor currents of DFIM with variable torque command.

very small it is because of variable torque command, but practically rotor speed almost constant.

The predictive DTC strategy has good tracking behavior and it is confirmed that the reduction in torque and flux ripples is achieved as there are absolute absence of over currents and reduced ripples in rotor currents as shown in **Figure 4(f)**. From the **Figure 4(f)**, it is noticed that there is continuous increase and decrease in the amplitude of rotor currents, similar to stator currents as shown in **Figure 4(a)**.

5.3 Performance of DFIM near synchronous speed

In this Section, the performance of the proposed control scheme of DFIM has been investigated near the synchronous speed. This is examined by varying the speed of DFIM from 1580 rpm (hyper synchronous value) to 1340 rpm (sub synchronous value) in terms of sine wave with frequency of 3 rads^{-1} and phase shift of 90°, with the reference values of torque and rotor flux are set to 0.4 pu and 1 pu respectively.

Even when the speed command is varied suddenly from hyper synchronous value to sub synchronous value no much transient peaks occur in stator currents, as shown in **Figure 5(a)**, which clearly emphasizes there is reduction in over currents in stator and this is because of proper selection of switching sequence of rotor voltage vectors.

As shown in **Figure 5(b)**, the torque developed by the machine for proposed strategy and classical DTC and it closely follows the reference torque which means the dynamic performance of the machine is quite satisfactory but when the rotor speed nears the synchronous speed at around 0.5 s, variable torque ripple is produced. This variability in the torque ripple is due to continuous selection of zero voltage vectors at that instant. It indicates, the smaller amplitude of rotor voltage vector is required at the instant of rotor speed nearing the synchronism, which

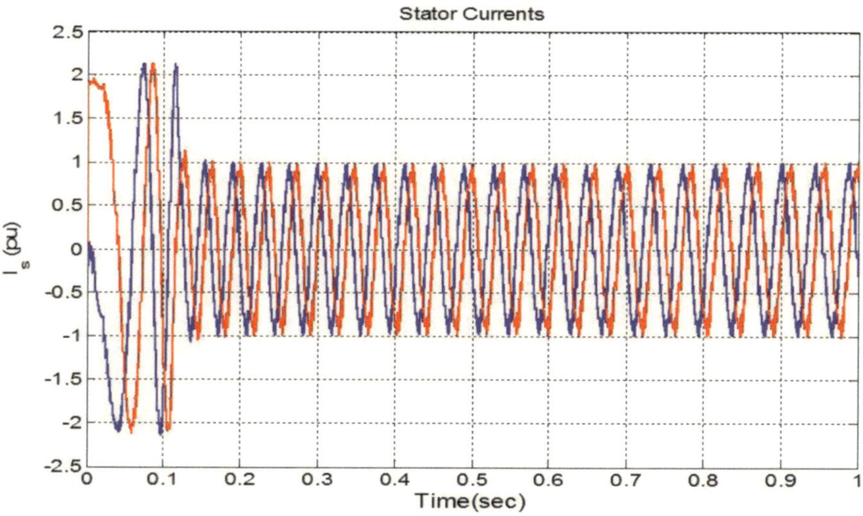
Stator Currents
I_s (pu)
Time(sec)

(a)

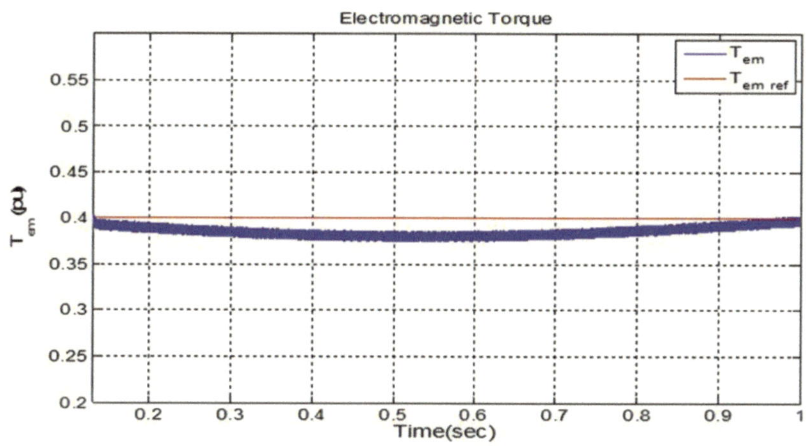
Electromagnetic Torque
T_em
T_em ref
T_em (pu)
Time(sec)

(i)

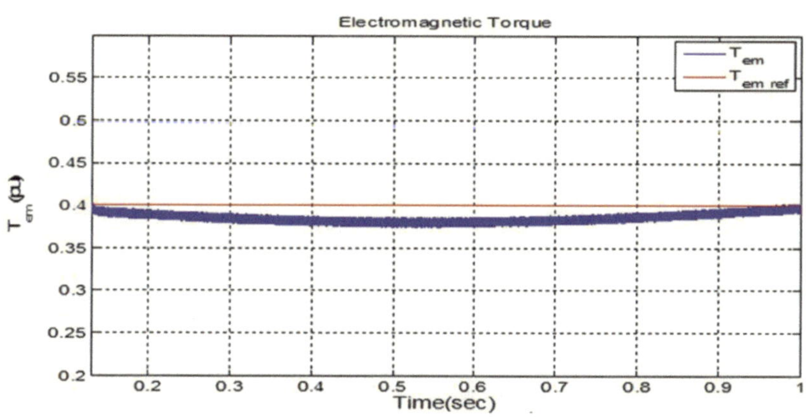
Electromagnetic Torque
T_em
T_em ref
T_em (pu)
Time(sec)

(ii)

(b)

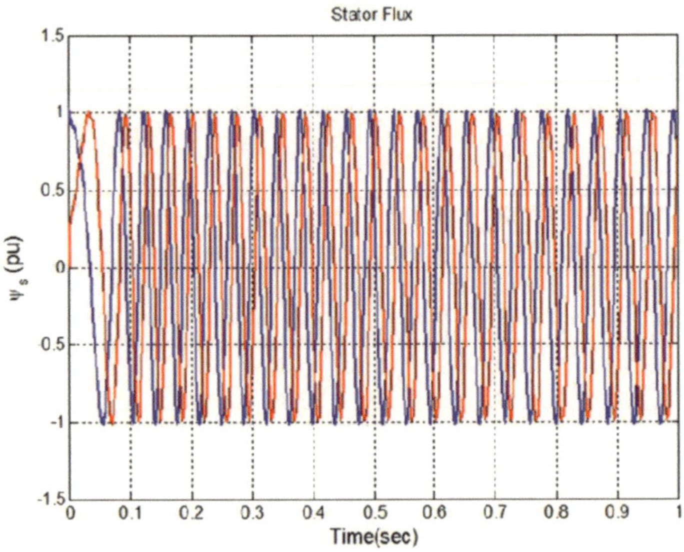
Stator Flux
ψs (pu)
Time(sec)

(c)

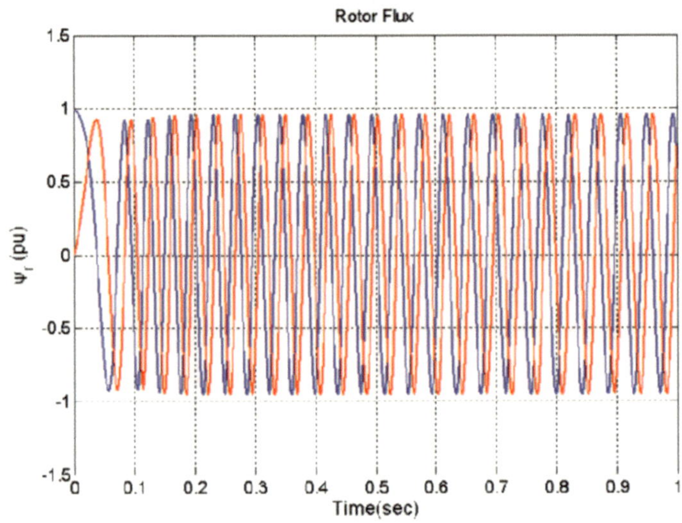
Rotor Flux
ψr (pu)
Time(sec)

(d)

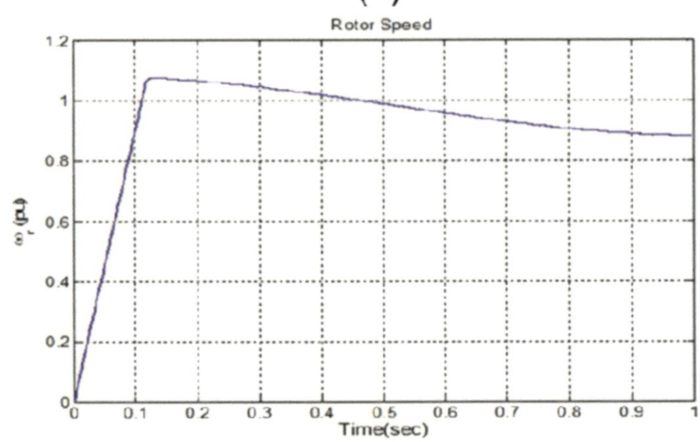
Rotor Speed
ωr (pu)
Time(sec)

(e)

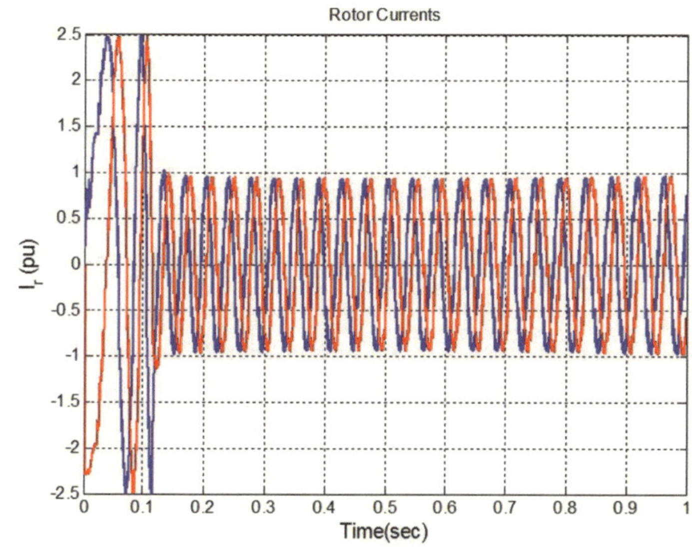

(f)

Figure 5.
(a) Response of stator currents of DFIM with variation in rotor speed from 1580 to 1340 rpm. (b) Torque response of DFIM with variation in rotor speed from 1580 to 1340 rpm (i) proposed strategy and (ii) classical DTC. (c) Response of stator flux of DFIM with variation in rotor speed from 1580 to 1340 rpm. (d) Response of rotor flux of DFIM with variation in rotor speed from 1580 to 1340 rpm. (e) Rotor speed of DFIM with variation in rotor speed from 1580 to 1340 rpm. (f) Response of rotor currents of DFIM with variation in rotor speed from 1580 to 1340 rpm.

actually causes the reduction in torque ripple and this leads to degradation of quality of control. As shown in **Figure 5(b)**, the ripples are reduced in electromagnetic torque response.

Stator flux responses are shown in **Figure 5(c)**. From the figure, it is observed that the stator flux remains constant and is not affected by change in rotor speed and also it is sinusoidal in nature.

The response of rotor flux is shown in **Figure 5(d)**, from the figure, it is observed that there is no effect on the rotor flux due to sudden change in the rotor speed and rotor flux is also not affected by the changeover.

The response of rotor speed is as shown in **Figure 5(e)**, when reference speed is varied from hyper synchronous value to sub-synchronous value and it is observed that the rotor speed response of DFIM follows the command speed.

Similar to stator currents as shown in **Figure 5(a)**, no much transient peaks occur in rotor currents, as shown in **Figure 5(f)**, that is, there is reduction in over currents in the rotor, which is because of proper selection of switching sequence of rotor voltage vectors.

6. Conclusions

The proposed control method makes two general contributions to the predictive control techniques. Firstly, it shows that using instead of two voltage vectors operating three appropriate vectors, allows operating at low constant switching frequency. Secondly, it is crucial to achieve the whole good performance of the DFIM, in terms of torque and current ripples by reducing the ripples of both directly controlled variables instead of only one.

From the proposed control method, it is possible to reduce the stress of the switching devices of the voltage source converter, in terms of low constant switching frequency behavior and switching power losses reduction, often demanded requirements in high power applications.

It presents good tracking behavior, capable of working at variable speed operation conditions, for both motoring and generating modes at sub- synchronous and hyper synchronous speeds when compared to DTC technique, making this control suitable for applications such as wind power generation.

The new DTC technique allows obtaining quick dynamic responses in respect to DTC method, with absolute absence of non-desired over currents in the machine. It ensures reduced torque and flux ripples, due to the control effect. The simulation results showed the effectiveness of the proposed method, to control the torque and the flux of the DFIM at considerably low constant switching frequency.

7. Summary

In this Chapter, new predictive DTC has been developed for DFIM. The proposed control scheme uses two voltage vectors instead of three voltage vectors and it allows operating at low constant switching frequency and reduces torque and flux ripples, and also capable of working at variable speed operating conditions for both motoring and generating modes at sub-synchronous and hyper synchronous speeds compared to classical DTC technique. The comparison of torque and flux ripple values (difference of maximum to minimum ripple value) and its reduction given by the difference of maximum value and minimum value to average value is given in the **Table 5** below.

S. No.	Parameter	Classical DTC scheme	Predictive DTC scheme
1	Torque ripple	0.3 pu	0.01 pu
2	Reduction in torque ripple	2.5%	1.25%
3	Rotor flux ripple	0.018 pu	0.01 pu
4	Reduction in rotor flux ripple	1.8%	1.05%

Table 5.
Torque and flux ripple reduction comparison.

Author details

Gopala Venu Madhav[1*] and Y.P. Obulesu[2]

1 Department of EEE, Anurag Group of Institutions, Ghatkesar, TS, India

2 School of Electrical Engineering, VIT University, Vellore, Tamilnadu, India

*Address all correspondence to: venumadhav.gopala@gmail.com

IntechOpen

© 2019 The Author(s). Licensee IntechOpen. This chapter is distributed under the terms of the Creative Commons Attribution License (http://creativecommons.org/licenses/by/3.0), which permits unrestricted use, distribution, and reproduction in any medium, provided the original work is properly cited.

References

[1] Blaschke F. A new method for the structural decoupling of a.C. induction machines. In: Proc. IFAC Conf., Dusseldorf, Germany, October 1971. 1971. pp. 1-15

[2] Takahashi I, Ohmori Y. High-performance direct torque control of an induction motor. IEEE Transactions on Industry Applications. 1989;**IA-25**(2): 257-264

[3] Depenbrock M. Direct self-control (DSC) of inverter-fed induction machine. IEEE Transactions on Power Electronics. 1988;**PE-3**(4):420-429

[4] Flach E, Hoffmann R, Mutschler P. Direct mean torque control of an induction motor. In: Proc. EPE'97 Conf. 1997. pp. 672-677

[5] Casadei D, Serra G, Tani K. Implementation of a direct control algorithm for induction motors based on discrete space vector modulation. IEEE Transactions on Power Electronics. 2000;**15**(4):769-777

[6] Kang JK, Sul SK. New direct torque control of induction motor for minimum torque ripple and constant switching frequency. IEEE Transactions on Industry Applications. 1999;**35**(5): 1076-1086

[7] Martins CA, Roboam X, Meynard TA, Carvalho AS. Switching frequency imposition and ripple reduction in DTC drives by using a multilevel converter. IEEE Transactions on Power Electronics. 2002;**17**(2): 286-297

[8] Sarasola I, Poza J, Rodríguez MA, Abad G. Predictive direct torque control of brushless doubly fed machine with reduced torque ripple at constant switching frequency. In: Proc. IEEE ISIE'07 Conf. 2007. pp. 1074-1079

[9] Aurtenechea S, Rodríguez MA, Oyarbide E, Torrealday JR. Predictive control strategy for dc/ac converters based on direct power control. IEEE Transactions on Industrial Electronics. 2007;**54**(3):1261-1271

[10] Pena R, Clare JC, Asher GM. Doubly fed induction generator using back-to-back PWM converters and its application to variable speed wind-energy generation. IEE Proceedings—Electric Power Applications. 1996;**143**: 231-241

[11] Gomez SA, Amenedo JLR. Grid synchronization of doubly fed induction generators using direct torque control. In: Proc. IEEE IECON'02 Conf. 2002. pp. 3338-3343

[12] Datta R, Ranganathan VT. Direct power control of grid-connected wound rotor induction machine without rotor position sensors. IEEE Transactions on Power Electronics. 2001;**16**(3):390-399

[13] Abad G, Rodríguez MA, Poza J. Predictive direct power control of the doubly fed induction machine with reduced power ripple at low constant switching frequency. In: Proc. IEEE ISIE'07 Conf. 2007. pp. 1119-1124

Chapter 6

Study of the Parameters of the Planner with a Screw Working Body

Juraev Tojiddin Khayrullaevich,
Norov Sobirjon Negmurodovich
and Musulmanov Furqat Shodiyevich

Abstract

This chapter examined the theoretical background of the use of a screw working body in front of the planner bucket and conducting experiments in laboratory conditions with the proposed working body. This work supports the practical solution of using a screw working particle in the current field planning. Significance of the work reducing traction resistance to soil movement up to 20% enables the tractor unit to work at higher speeds of translational motion; the latter contributes to increased productivity, improved planning quality and reduced cash costs per unit of work performed. The chapter was prepared under results of research in the Mechanics Laboratory of Bukhara Engineering Technological Institute.

Keywords: productivity of the screw working body, translational speed of the planning unit, screw pitch, screw diameter, drawing prisms

1. Introduction

Further development of agriculture in modern conditions determines the introduction of new advanced technologies and machines for their implementation.

As we know, for increase of volume agricultural products, it is need to intensification of agricultural production, one of the means of which is wide land reclamation, which provides, as one of the most important measures, the planning of the irrigated land surface. Alignment of the surface of the fields by the long-base planners is of great importance in the range of planning works, and research work to improve these planners is carried out both in Uzbekistan and abroad.

The analysis of the state of the issue showed the need to improve the quality of planning work, increase the productivity of long-base planners and increase the levelling ability of the latter [1, 2].

2. Literature review

By world and domestic farming practices, it has been proved that planning or levelling the surface of the fields is the main land reclamation measure designed

IntechOpen

to eliminate irregularities on the field under sowing in the form of various rise and falls.

Many domestic and foreign scientific works [3–6] have been published on the need for field planning, authors of which have been examined the issue from different perspectives and pointed out the advantages of this operation. Moreover, there are many advantages, whether it is capital and operational planning or micro alignment. Experiments have established that under conditions of turbulent terrain, crop losses amount to 40% of the potential crop. Cotton grown on mounds and lowlands gives raw cotton of lower quality: fibre strength, grade, ripeness, etc. are reduced.

In areas with saline soil, a high-quality planning prevents the formation of saline spots (areas) and thereby ensures sustainable yields of all types of crops [6, 7]. Flushing saline lands without planning are ineffective.

According to Ref. [6], it was established that uniform soil moisture is achieved on the planned field, and the irrigation water consumption is significantly reduced.

In the United States, great importance is given to the planning of the surface of irrigated areas. Annually, in this country, approximately more than half of the irrigated areas are subject to planning [3, 6, 7]. Americans do not spare the cost of land planning, as this increases income from irrigated land and reduces the cost of production.

To ensure the high quality of the technological processes, including irrigation, it is necessary to pay special attention to the capital planning and the mandatory periodic implementation of the field operational planning.

3. Research methods

It is known that the magnitude and the nature of the change in the angular speed of a material particle determine the productivity and energy indicators for conveying the material with a screw [2, 8].

Let us consider the movement of a material particle of mass d_m located at point O of an inclined cylindrical screw at a distance r from the axis of the screw (**Figures 1** and 2) and moving along the trajectory of the absolute movement AB; the axes τ, b and n are, respectively, tangent, binormal and normal to the trajectory of absolute motion [9]. The n axis is directed towards the centre of curvature and

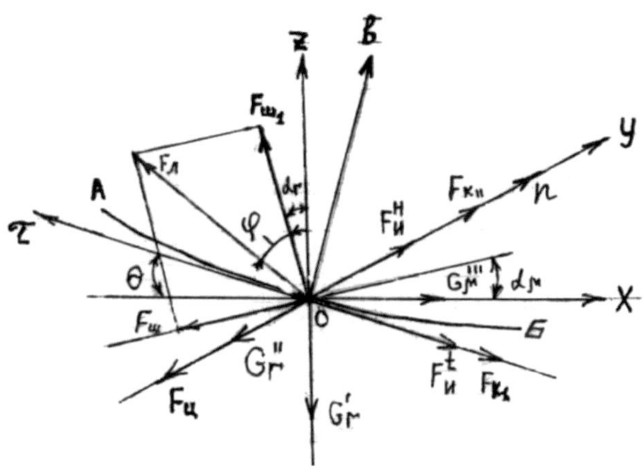

Figure 1.
Movement of a material particle of mass d_m located at point O.

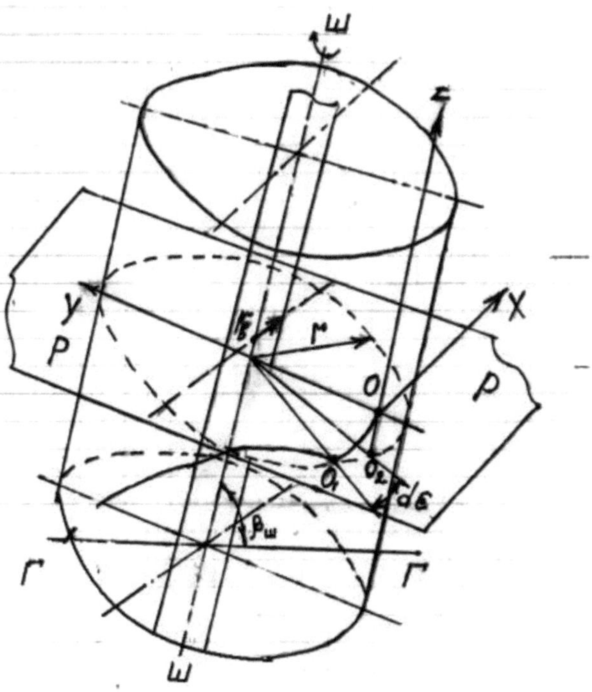

Figure 2.
Material particle of mass d_m located at point O of an inclined cylindrical screw at a distance r.

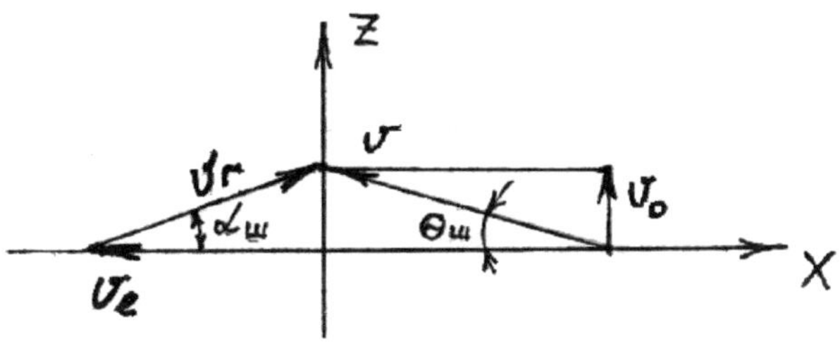

Figure 3.
Particle's absolute speed vector $\overline{V}$.

coincides with the y axis. The Z axis is parallel to the axis of the screw, and the x and y axes are located on tangent and normal in plane PP of the screw, perpendicular to its axis ШШ; GG horizontal plane.

The following forces impact on the soil particle: particle gravity Gr = gdm, which can be divided into three components, axial Gr^I = Gr.sin $\beta_{Ш}$ (along the z axis), radial G_r^{II} = G_r.cos $\beta_{Ш}$ cosε (along the y axis) and tangent G_r^{III} = G_r.cos$\beta_{Ш}$·sinε (along the x axis); centrifugal force $F_ц = \omega_r{}^2 \cdot \mu \cdot d \cdot m$ (along the y axis), friction force of a particle on a blade casing F_{k_1} (along the τ axis), and friction force on a helical surface $F_{Ш}$ (at an angle of inclination of the helix α_r to the x axis); tangential inertial force, $F_u{}^t$, acting along the tangent to the trajectory of the absolute motion of the particle (τ axis) and directed opposite to the particle's absolute speed vector $\overline{V}$ (**Figure 3**); normal inertial force directed to the centre of curvature of the trajectory (along the n axis); and normal reaction of the adjacent layer $F_{K_{II}}$ (along the n axis) and the helical surface $F_{Ш_I}$ (at an angle a_r to the z axis). $\beta_ш$ is the angle of inclination of the screw to the horizon; and ε is the current angle of rotation of the particle, measured from the projection of the O_2 particle on the PP plane. The resultant $F_л$ of the normal reaction of the helical surface $F_{\text{ш}_I}$ and the friction force against the helical surface $F_{ш} = fF_{ш_1}$ is deviated from the normal and the helical surface by the angle of friction $\phi = arctgf$, where f is the coefficient of friction of the soil over the screw metal. If we consider that the loosened soil before the bucket of the planner is clay, then the value of this coefficient is 0.6 ... 0.7. [8]

The friction force of a particle on a bucket caused by the combined action of forces $F_{ц}$ and $G^{II}{}_{r}$ is equal to:

$$F_{k_1} = f_r(F_ц + G_r{}^{II}) = f_r(\omega_r{}^2 \cdot r + g\cos\beta_ш \cdot \cos\varepsilon) \cdot dm,$$

where f_r and f, respectively, are the friction coefficients of the particle on the bucket and the adjacent layer of material and the helical surface.

Absolute particle speed:

$$\vartheta = \sqrt{\vartheta_t{}^2 + \vartheta_o{}^2} = r\sqrt{\omega_r{}^2 + (\omega - \omega_r)^2 \cdot tg^2 a_r} \tag{1}$$

where ϑ_t is tangential particle speed at the radius r from the axis of the screw, $\vartheta_t = \omega_r \cdot r$; ϑ_o is the axial speed of the particle at a radius r from the axis of the screw, $\vartheta_o = (\omega \cdot \omega_r)tga_r$; ω is screw angular speed; and a_r is the angle of inclination of the helix of the screw on the radius r (**Figure 2**).

The tangent force of inertia is defined as follows:

$$F_u{}^t = \frac{d\vartheta}{dt}dm = \frac{r[\omega_r - (\omega - \omega_r)tg^2 a_r]}{\sqrt{\omega_r{}^2 + (\omega - \omega_r)^2 tg^2 a_r}} \cdot \frac{d\omega_r}{dt}dm \tag{2}$$

Normal inertial force:

$$F_u{}^н = \vartheta^2 \cdot r_a{}^{-1} \cdot dm = r^2\left[\omega_r{}^2 + (\omega - \omega_r)^2 \cdot tg^2 a_r\right] \cdot \left[r(1 + tg^2\theta)\right]^{-1}, \tag{3}$$

where r_a is the radius of curvature of the trajectory at the selected point, $r_a = r(1 + tg^2\theta)$; and θ is the angle of inclination of the helix of the particle trajectory to the axis X (**Figure 2**)

$$tg\theta = tga_ш(\omega - \omega_r)\omega_r{}^{-1}, \tag{4}$$

where $a_{\text{ш}}$ is screw helix angle at the periphery.

According to the D'Alembert's principle [8], the equation of dynamic equilibrium of a material particle in the projections on the axis of the natural trihedral of the trajectory (τ, b, n)(**Figure 2**) will be

$$\begin{aligned}\sum\tau = {} & \left[F_л \sin(a_r + \theta + \phi) - G_r{}^{III}\cos\theta - F_{k_1} - F_u{}^t - G_r{}^1\sin\theta\right] \cdot \cos\varepsilon - \\ & -\left(F_ц + G_r{}^{II} - F_u{}^H - F_{k^{II}}\right) \cdot \sin\varepsilon = O,\end{aligned} \tag{5}$$

$$\sum b = \pm F_л \cdot cos(a_r + \theta - \phi) + G_r{}^{III} \cdot \sin\theta - G_r{}^1 \cdot \cos\theta = 0, \tag{6}$$

$$\begin{aligned}\sum n = {} & \left(F_ц + G_r{}^{II} + F_u{}^H - F_{k_{ii}}\right) \cdot \cos\varepsilon + [F_л \sin(a_r + \theta + \phi) - \\ & -G_r{}^{III} \cdot \cos\theta - F_{k1} - F_u{}^t - G_r{}^1\sin\theta] \cdot \sin\varepsilon\end{aligned} \tag{7}$$

Solving these equations jointly, excluding the force $F_л$ from them, after the corresponding transformation of the exception and time by expressing the elementary angle of rotation along the arc O_1, O_2 (**Figure 1**) of the particle $d\varepsilon = (\omega - \omega_r)dt$, we obtain:

$$\frac{d\omega_r}{d\varepsilon} = \frac{\pm\dfrac{sin([a_r + \phi)\omega_r + \cos(a_r + \phi) \cdot tga_r(\omega - \omega_r)}{\cos(a_r + \phi)\omega_r - \sin(a_r + \phi)tga_r(\omega - \omega_r)}}{(\omega - \omega_r)r[\omega_r - (\omega - \omega_r)tg^2 a_r]} \rightarrow$$
$$\rightarrow \frac{[g \cdot \sin\beta_ш \cdot \omega_r - g \cdot \cos\beta_ш \sin\varepsilon \cdot tga_r(\omega - \omega_r)] -}{(\omega - \omega_r)r[\omega_r - (\omega - \omega_r)tg^2 a_r]} \rightarrow$$
$$\rightarrow \frac{-f_r(\omega_r^2 + g \cdot \cos\beta_ш \cdot \cos\varepsilon)x\sqrt{\omega_r^2 + (\omega - \omega_r)^2 \cdot tg^2 a_r} -}{(\omega - \omega_r)r[\omega_r - (\omega - \omega_r)tg^2 a_r]} \rightarrow \tag{8}$$
$$\rightarrow \frac{-g \cdot \sin\beta_ш \cdot tga_r(\omega - \omega_r) - g \cdot \cos\beta_ш \sin\varepsilon \cdot \omega_r}{(\omega - \omega_r)r[\omega_r - (\omega - \omega_r)tg^2 a_r]}$$

By integrating Eq. (8) by the Euler method [3], we can obtain the curves of the dependence of ω_r on ε for screws with different parameters.

In the inclined screw, there is a periodically steady motion of the material particle. The maximum value of ω_r^{max} is on the plot with the values of the angle $2k\pi > \varepsilon > (2k - 1)\pi$, and the value of ω_r^{min} is in the zone $(2k + 1)\pi > \varepsilon > 2k\pi$, where (k) is any number. In order to simplification for inclined screws, it is possible to take the average value of the angular speed in the area of periodically steady motion:

$$\omega_r^{cp} = \int_{\varepsilon=2k\pi}^{\varepsilon=(2k+1)\pi} \cdot\omega_r d\varepsilon \cong \frac{\omega_r^{2k\pi} + \omega_r^{(2k+1)\pi}}{2} \tag{9}$$

The values of ω_r^{max} and ω_r^{min} can be found by progressive approximation from Eq. (8), equating the right side to zero at $\varepsilon = 2k\pi$ and $\varepsilon = (2k + 1)\pi$. However, for practical purposes, the angular speed of a soil particle at the periphery of the screw can be taken as:

$$\text{for vertical screws, } 0,4\omega < \omega_{r_ш} < 0,5\omega,$$
$$\text{for steeply inclined screws, } (\beta_ш \geq 30^o)\ 0,3\omega < \omega_{r_ш} < 0,4\omega,$$

where $\omega_{r_ш}$ is the particle angular speed at the periphery of the screw, $\sec^{-1}$.

The study of Eq. (8) showed that in a certain zone limited by the cylindrical contour of the soil shaft, soil particles relatively quickly acquire the angular speed of the screw and their lifting stops [8]. The value of the so-called critical radius $r_{кр}$ depends on the initial conditions and is determined from Eq. (8) after substituting $\omega_r = \omega$ at $d\omega_r/d\varepsilon = 0$.

In the horizontal ($\beta_ш < 30°$) screw, during the period of steady motion, the angular speed of the particle is zero. The angle of rotation of the particle ε, at which the steady motion begins, depends on the initial conditions and can be found from Eq. (8):

$$\varepsilon = arctg[f_r \sin(a_r + \phi)\cos^{-1}(a_r + \phi)] \tag{10}$$

Maximum productivity on loosened soil, determined by the throughput between the upper turns of the screw will be:

$$\Pi = \int_{r_{кр}}^{r_{ш}} \vartheta_r dS, \tag{11}$$

where $r_{ш}$ is the outer radius of the screw; ϑ_r is the ground sliding speed on the helical surface of the screw (relative speed); and dS is the elementary cross-sectional area of the soil located between the upper turns in a plane perpendicular to the relative speed vector.

The screw capacity in a dense body will be:

$$\Pi_T = \frac{1}{K_p} \int_{\kappa_{кр}}^{r_{ш}} \vartheta_r \cdot dS^I, \tag{12}$$

where κ_p is soil loosening coefficient; for our case, κ_p = 1.14 ... 1.28 [8].

Soil in the cross section of the screw by a plane, passing through the axis of the screw, occupies an area bounded from below by a straight-line perpendicular to the axis of the screw (**Figure 4**). Then in the cross section, we get a rectangle of length l^{II}, which can be taken equal to the pitch of the screw $l^{II} = l^{I}_{III}$.

The elementary area of soil cross-section at a distance r from the screw axis will be equal to:

$$dS = dS_1 \cdot \cos a_r = l^1{}_{ш} d_r \cos a_r = 2\pi l^1{}_{ш} \left(\sqrt{4\pi^2 \cdot r^2 + \left(l^1{}_{ш}\right)^2} \right)^{-1} \cdot rd_r, \tag{13}$$

where dS_1 is the elementary area of soil in axial cross-section (**Figure 4**).

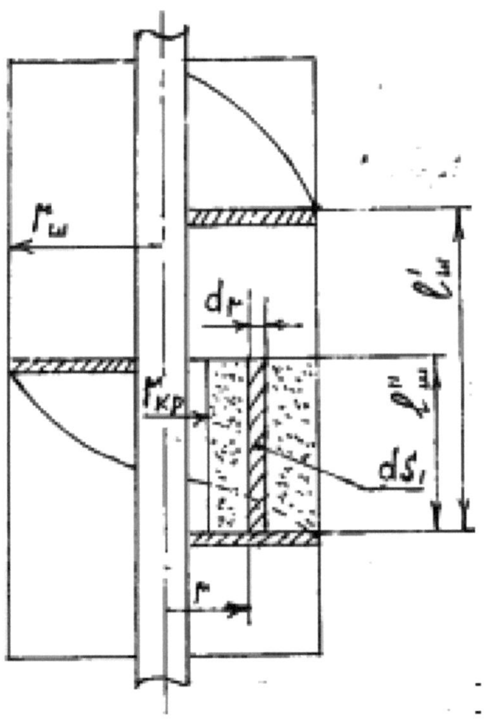

Figure 4.
Position of soil area in the cross section of the screw.

In horizontal and gently dipping screws at inclination angle $\beta_{ш} \leq 30°$, the angular speed of the material particle is $\omega^{cp}{}_{r_{ш}} = 0$. To calculate the capacity (m^3/h), we can use [1] the following formula:

$$П_T = 450\left(d_{ш}{}^2 - d_b{}^2\right)l_{ш}{}^1\omega \cdot K_H K_\beta \cdot K_p{}^{-1}, \tag{14}$$

where $d_{ш}$, $d_{в}$, respectively, are the diameters of the screw and shaft, m; $K_н$ is the screw filling coefficient, for our case we can take equal to $K_н = 0{,}2 \ldots 0{,}4$; and K_β is the coefficient taking into account the angle of inclination of the screw to the horizon $\beta_{ш}$, $K_\beta = 1{,}0 \ldots 0{,}8$ [8].

3.1 Selection and justification of the main parameters of the screw

The main parameters of working elements of the screw include the length of the conveying part—$l_{ш}$, the length of the cutting part—l_p, the diameter of the screw—$d_{ш}$, peripheral speed on the cutting edge—$\vartheta_{окр}$, the pitch of the screw—$l^1{}_{ш}$ and the working speed of movement—ϑ_p.

The length of the conveying and cutting parts of the screw is taken constructively, based on the type of screw and the parameters of the medium being processed. For preliminary calculations, we can take the length of the conveying part of the horizontally located screw $l_{ш} = l_p = (0{,}7 \ldots 0{,}8)\ в_n$, where $в_п$ is the width of the planner scoop.

The diameter of the screw $d_{ш}$ with a horizontal working body at the given productivity $П_т{}^1 = П_т$ can be determined from formula (14) after some transformations, m;

$$d_{ш} \geq \sqrt{П_T{}^1 \cdot K_p\left(900\vartheta_{окр} \cdot K_a K_н \cdot K_\beta\right)^{-1} + d^2{}_{в}}, \tag{15}$$

where $\vartheta_{окр}$ is the peripheral speed on the cutting edge of the screw, $\vartheta_{окр} = 1.5 \ldots 3$ m/sec; and K_a is the coefficient taking into account the inclination of the cutting edge of the screw, $K_a = l^1{}_{ш}/d_{ш} = 0{,}7 \ldots 1{,}0$. Other designations see from the formula (14).

The step of the horizontal screw $l^1{}_{ш}$ is taken equal to $l^1{}_{ш} = K_a d_{ш}$, the value of K_a is taken depending on the inclination of the cutting edge of the screw. For our case, we can take K_a equal to 0.85.

The working speed of the soil movement with the screw should be equal to the speed of filling the planner bucket with soil. The latter depends on the forward speed of movement of the planner. For our case, with a certain accuracy, we can take $\vartheta_{гр} = \vartheta_{ков} = \vartheta_п$, where $\vartheta_{гр}$ is the speed of the soil movement with the screw, $\vartheta_{ков}$ is the speed of filling the planner bucket with soil and $\vartheta_п$ is the forward speed of the planner, m/s.

The working speed of the planner's movement can also be determined from the conditions for ensuring a given efficiency on a cut of soil with the planner's bucket. For a horizontal screw, operating speed ($\vartheta_п$) of movement is (m/s):

$$\vartheta_П = П_T \cdot l_p{}^{-1} \cdot h_p{}^{-1}, \tag{16}$$

where $П_т$ is the efficiency of the planning unit on cut of soil, m^3/sec; l_p is the cutting length of the planner's knife, m; and h_p is the thickness of the cut soil layer, m.

Based on the analysis of the above theoretical background in the determination of the efficiency of the screw working body, it is suggested that with the increase in the speed of rotation and in diameters of the screw, the productivity of the screw working body increases. The screw pitch is also of great importance, with an increase in which the volume of soil moved to the sidewalls of the planner bucket increases, which in turn contributes to an even distribution of the soil of the drawing prism along the width of the planner passage. With an increase in the speed of forward movement of the planner increases the working capacity of the screw working body, that is screws move a large volume of soil to the sides relative to each other. However, such an improvement in the work of working element of the screw for our case, as shown by selective experiments with the experimental sample of a mini-planner, occurs up to a speed of 2 m/s of the forward movement of the unit. Above this speed, the screws begin to become clogged with soil, and the technological process of the screw working body is violated.

We derived the equation of traction resistance of the planner, which has the following form:

$$P = f_n G_n + \frac{\left[\sin\beta(1-f^2)2f\cos\beta\right]\tau\cdot S}{\sin(\beta+\theta)(1-\mu f)+\cos(\beta+\theta)(\mu+f)} + \frac{1}{3}\lambda_v l^3 tg^2\varphi_0 tg^2\left(45-\frac{\beta}{2}\right)f + \\ +G_{np}\left(f\cos^2\beta + tg\rho + \frac{2V^2\sin^2\frac{\beta}{2}}{Kycm\cdot g}\right) \quad (17)$$

where f_n– planner rolling resistance coefficient; G_n– mass of the planner; β– cutting angle; f, μ– coefficients of friction of soil on steel and soil on soil; τ– shear stress; S– shear area; θ– shear angle; λ_v– volumetric mass of soil as in the function of movement speed; l– bucket side length; φ_0– angle of slope of the soil roller during movement; G_{np}– mass of the soil roller; V– unit movement speed; and $Kycm$– coefficient taking into account the design of the rear wall of the bucket [10].

4. Results and discussion

From the analysis and conclusions we can see that with the increase in the rotation speed and diameter of the screw, the productivity of the screw working element increases. The screw pitch is also of great importance, with an increase in which increases the amount of soil movement to the sidewalls of the planner bucket, which in turn contributes to uniform distribution of the soil of the drawing prism along the width of the planner. With the increase in the speed of translational movement of the planner increases the functionality of the screw working element, that is the screws move a large volume of soil to the sides relative to each other. Below in **Figure 5**, the location of the screws in the bucket of the planner is shown.

However, such improvement in the work of the screw working element for our case, as shown by selective experiments with the experimental sample of a mini-planner, occurs up to the speed of 2 m/s of translational movement of the unit. Above this speed, the screws begin to be clogged with soil, and the technological process of the screw working body is violated. Below in **Figure 6**, experimental sample of a mini-planner with a screw working body is shown.

The above analysis requires investigating the productivity of the screw working body, depending on the rotation speed, diameter and pitch of the screw. Therein, the translational speed of the planning unit is also of significant importance. Because, the volume of the planner's bucket filled with soil per unit of time should be equal to the volume of the processed soil by the screws.

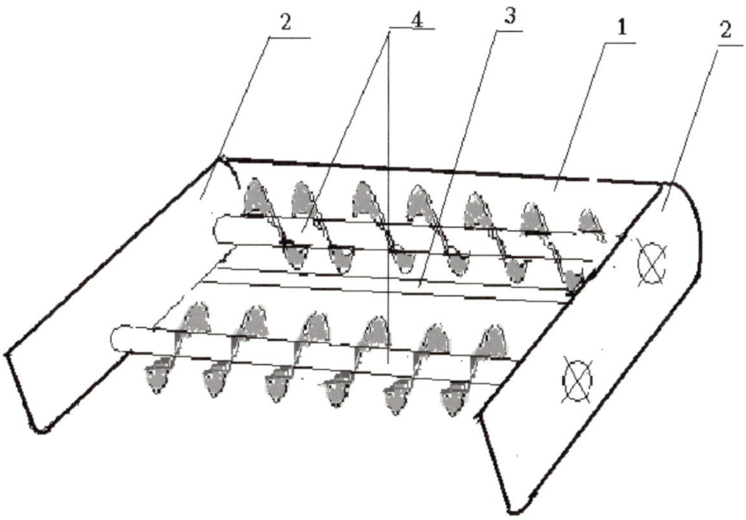

Figure 5.
Scheme of the location of the screws in the bucket of the planner (1—rear wall of the bucket, 2—tank walls, 3—knife of the rear wall and 4—screws).

Figure 6.
General view of the experimental prototype of the mini planner.

Below are the curves (**Figures 7–9**) of the change in productivity of the screw working element of the planner depending on the rotation speed, diameter and pitch of the screw.

As we can see from the graph (**Figure 7**), with the increase in screw rotation, the screw productivity is directly proportional to screw rotation. At a screw rotation of 40 rpm, the screw capacity is 2.56 m^3/h, and at screw rotation of 240 rpm, the screw capacity increases to 15.38 m^3/h. That is, the productivity of the screw increases by six times. This increase in productivity approximately corresponds to the desired performance of the planning unit.

The change in the performance of the screw in the function (D_s) of the diameter of the screw is shown in **Figure 8**.

Analysing this graph, we can say that the change in productivity along its diameter has curvilinear nature. Moreover, part of the curve to the point corresponding to D_s = 180 mm is a power-law nature, and then the curve changes linearly. Further increase in the diameter of the screw (D_s) leads to the increase in its productivity so that the planner will not be able to provide the screws with soil for their normal operation. In addition, the increase in the given productivity of the

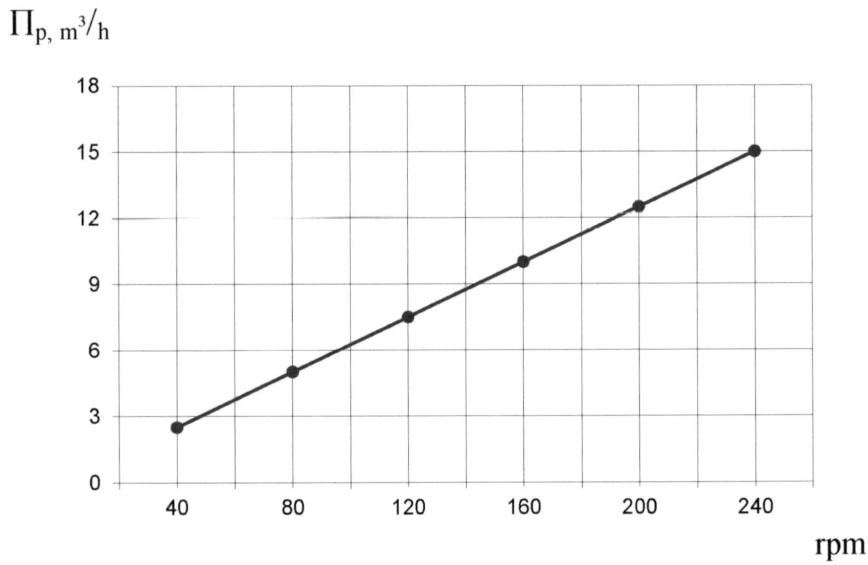

Figure 7.
Change of the productivity of the screw working element depending on the speed of rotation of the screws (at D_p = 0.18 m, S_p = 0.15 m and $K_н$ = 0.28—bucket filling ratio).

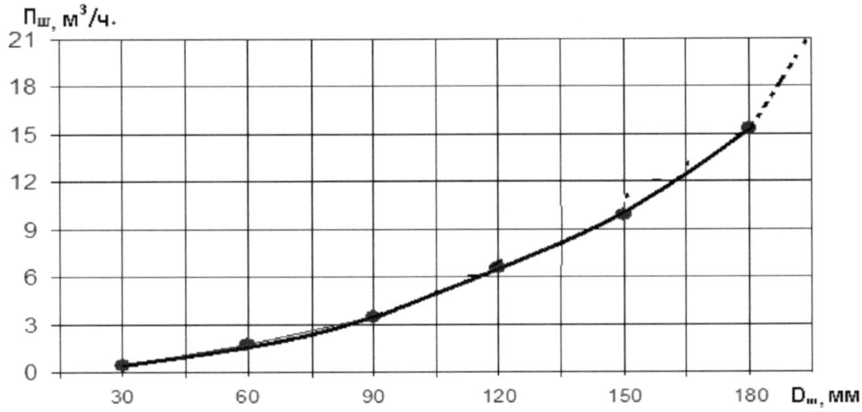

Figure 8.
Change in productivity of the screw working element depending on the diameter (Ds) of the screw (at S_s = 0.15 m, n_s = 240 rpm and $K_н$ = 0.28).

scheduler, which is accompanied by the increase in the translational speed of the movement of the unit above 7.5 km/h, leads to a disruption of the technological process of planning and the decrease in the quality indicators of field levelness [3].

The graph of the performance of the screw working element depending on the pitch (S_m) of the screw is shown in **Figure 9**.

As we can see from the graph, with the increase in the pitch of the screw (S_s), its productivity changes in direct proportion to the changes in the pitch (S_s), that is the functional change in the curve is linear. If with a screw pitch of 30 mm, the screw capacity is 3.07 m^3/h, then with a screw pitch of 180 mm, the productivity increases to 18.45 m^3/h, that is it increases six times.

Analysis of the graph (**Figure 9**) shows that increasing the screw pitch to 180 mm increases the capacity to 18.45 m^3/h, which is almost to 3 m^3 more than at the screw pitch S_s = 150 mm. Such an increase in the productivity of the screw

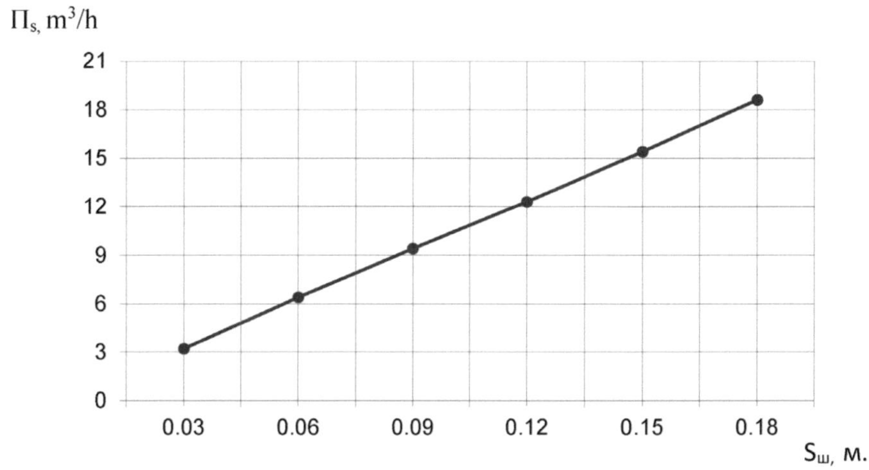

Figure 9.
Change in the productivity of the screw working element depending on the pitch (S_s) of the screw (at D_s = 0.18 m, n_s = 240 rpm and $K_н$ = 0.28).

working element could be obtained with a further increase in its diameter (D_s). But it is known that the increase in the diameter (D_s) of the screw is accompanied by greater energy consumption during its operation compared to the increase in the pitch of the screw. Therefore, a theoretical study of the work of the screw working element in the bucket of the planner allows us to conclude that for a given productivity of the planning unit, it is advantageous and advisable to use the screw parameters: D_s = 180 mm, n_s = 240 rpm and screw pitch S_s = 180 mm.

Increase in the productivity of the screw working element due to the increase in the pitch of the screw reduces the metal consumption of the screw and the corresponding material costs in comparison with the increase in the diameter (D_s) of the screw. In addition, further increase in the diameter of the screw causes difficulties in their layout in the bucket of the planner.

Using the methodology for conducting scientific research and processing the obtained data [2, 10], we derived empirical equations (formulas) of dependencies $y = f(n_ш)$, $y = f(D_ш)$ and $y = f(S_ш)$. For the graphs of **Figures 7–9**, respectively, $y_1 = 0,064X$; $y_2 = 474,7X^2$; and $y_3 = 102,5X$, the curve of which is consistent with the curves shown in **Figures 7–9**.

The screw working body surface was geometrically modelled by corresponding author and the offered geometric models give possibility to control engineering and technological parameters by optimization of geometric parameters of working surface, which may use this possibility in designing the working body of screw planners through exporting CAD models to CAE system [9, 11–15].

Author details

Juraev Tojiddin Khayrullaevich[1*], Norov Sobirjon Negmurodovich[2]
and Musulmanov Furqat Shodiyevich[1]

1 Tashkent Institute of Irrigation and Agriculture Mechanization Engineers, Bukhara Branch, Uzbekistan

2 Bukhara Engineering-Technological Institute, Uzbekistan

*Address all correspondence to: tojiddin_1968@mail.ru

IntechOpen

© 2020 The Author(s). Licensee IntechOpen. This chapter is distributed under the terms of the Creative Commons Attribution License (http://creativecommons.org/licenses/by/3.0), which permits unrestricted use, distribution, and reproduction in any medium, provided the original work is properly cited.

References

[1] Khasanov IS, Norov SN. Theoretical background of the results of experiments on the use of screw before the planner bucket. Universum: Technical Sciences Moscow. 2019;**12** (69):41–44

[2] Norov SN, Hamraeva MF. Analytical research of determining the maximum efficiency of planning machines. International Journal of Advanced Research in Science, Engineering and Technology. 2019:**6**(6):361-362. Available from: http://www.ijarset.com/upload/2019/june/37-IJARSET-Abrorov-51.pdf

[3] Norov SN. Research and Development of a Machine System for Complex Mechanization of Precultival Treatment of Agricultural Conditions in the Bukhara Region. Deutschland, Germany: LAP LAMBERT Academic Publishing; 2019. p. 65. Available from: https://www.lap-publishing.com/

[4] Efremov AN. Laser Planning of Irrigated Lands. Moscow. 2016. Available from: https://www.ya-fermer.ru/sites/default/files/lasernayaplanirovka_0.pdf

[5] Antonov EV. Development of technological process of planning rice drafts using the multifunctional planning machine [Dissertation for the degree of candidate of technical sciences]. Russia. 2008

[6] Efremov AN, Yu NS. Machines and technologies of capital planning of rice drafts. Tractors and Agricultural Machinery. 2016;**3**:37

[7] Efremov AN. The method of calculating the operational efficiency of levers. Construction and Road Machines. 2015;**5**:42-46

[8] Khasanov IS, Norov SN. Theoretical background for determining some parameters of the screw working jointly with the planner bucket. Agrarian Science Journal, Moscow, 2015;**6**

[9] Juraev TK. Design the working surface based on the particle trajectory. In: Proceeding Outs of Conference “AVIA-2013”. Kiev: NAU; 2013. pp. 21–23. Available from: https://scholar.google.com/scholar?cluster=7986884450713291378&hl=en&oi=scholarr

[10] Norov SN. Theoretical background for research of tractive effort of the planner. In: Collected papers of XXII International Scientific-Practical conference “Russian Science in the Modern World”. Moscow. 2019. Available from: http://xn–80aa3afkgvdfe5he.xn–p1ai/RNSM-22_originalmaket_N-1.pdf

[11] Juraev TK. Decision maintenance management problems in agriculture Engineering by Constructive Geometric Modeling Methods. Maintenance Management. Edited by Fausto Pedro García Márquez and Mayorkinos Papaelias. London, United Kingdom. 2020. pp. 23-37. https://mts.intechopen.com/storage/books/8623/authors_book/.

[12] Juraev T.Kh, Murodov N.M, Naimov S.T. Application the Geometric Modeling Methods and Systems in Design Engineering and Manufacturing on Example of Agriculture Engineering. /Design and Manufacturing. /Edited by Evren Yasa, Mohsen Mhadhbi and Eleonora Santecchia. First published in London, United Kingdom, 2020 by IntechOpen, 211-231 p. http://dx.doi.org/10.5772/intechopen.83290.

[13] Juraev TK. Creating the geometric database for product lifecycle management system in agricultural engineering. In: International Conference on Information Science and Communications Technologies ICISCT

2017 Applications, Trends and Opportunities; 2-4 November 2017, TUIT, Tashkent, Uzbekistan. IEEE Catalog Part Number: CFP17H74-CDR, ISBN: 978-1-5386-2167-7. 2017. Available from: https://www.researchgate.net/publication/321821311

[14] Juraev TK. Conceptual design of moldboard based on geometric modeling. In: Theory and Application of Design. Kiev: NAU; 2012. №. 1. pp. 75-81. Available from: http://www.irbis-nbuv.gov.ua/cgi-bin/irbis_nbuv/cgiirbis_64.exe?C21COM=2&I21DBN=UJRN&P21DBN=UJRN&IMAGE_FILE_DOWNLOAD=1&Image_file_name=PDF/tprd_2012_1_13.pdf

[15] Juraev TK. Modeling the directory curve of moldboard surface by conic. Applied Geometry and Engineering Graphic. Kiev: KNUCA; 2011. №. 87. Available from: https://scholar.google.com/scholar?cluster=2210113268218384828&hl=en&oi=scholarr

Chapter 7

Torque Ripple Reduction in DTC Induction Motor Drive

Adhavan Balashanmugham, Maheswaran Mockaisamy and Sathiyanathan Murugesan

Abstract

The asynchronous or Induction Motor (IM) is one of the most widely used electrical machines in the world, due to the three following advantages namely 1. Their construction is simple and rugged 2.The absence of slip rings, commutators and brushes make it cheaper, and 3.It is also maintenance free compared to DC motors and Synchronous motor due to wear and tear of brushes, slip rings and commutators respectively. The Section 1 deals with the introduction of induction motor and Direct Torque Control scheme. Section 2 briefly discusses the types of Induction motor. Section 3 tells about the control strategies of Induction motor respectively scalar control and vector control, and also briefly explains about Direct Torque Control (DTC) method. The Section 4 discuss about the Types of Control Strategies for Torque ripple Reductions in DTC as well as the two proposed schemes namely 1.Fuzzy Logic Controller (FLC) for DTC-SVM and 2.Artificial Neural Network (ANN) controller for DTC-SVM respectively for IM and its results, The two proposed schemes uses Hybrid Asymmetric Space Vector Pulse Width Modulation (HASVPWM) for switching the inverter. The Section 5 revels about the modern advanced techniques such as ANN and FLC based DTC.

Keywords: types of IM, control techniques of IM, direct torque and flux control (DTC), torque ripple reductions in DTC, modern strategies of DTC, fuzzy logic controller (FLC) for DTC-SVM, artificial neural network (ANN) controller for DTC-SVM, hybrid asymmetric space vector pulse width modulation (HASVPWM)

1. Introduction

The electric motors are electromechanical machines, which are used for the conversion of electrical energy into mechanical energy. The foremost categories of AC motors are asynchronous and synchronous motors. The asynchronous motors are called singly excited machines, that is, the stator windings are connected to AC supply whereas the rotor has no connection from the stator or to any other source of supply. The power is transferred from the stator to the rotor only by mutual induction, owing to which the asynchronous motors are called as induction machines. The induction motor is used widely in several industrial applications because of the following advantages 1. Ruggedness 2. Good efficiency 3. Simple and easy control. When the induction motor is compared to separately excited DC drives it is inferior because of coupled torque and flux. To bring high performance in induction motor drive the advanced control techniques of induction motor uses independent control of torque and flux, like in separately excited DC drives.

IntechOpen

The advanced control techniques such as field oriented control and direct torque control play vital role in today's high performance AC drives. Later in the 1980s, the direct torque control (DTC) method was proposed by Takahashi and Depenbrock [1, 2]. The direct torque control is a robust method compared to other methods. In this method by selecting optimum inverter switching modes the motor torque and flux are controlled independently and also direct. The primary input of the motor is stator voltage and stator current. From this the stator flux and electromagnetic torques are calculated. The torque errors and flux errors are limited within the hysteresis band. The Direct torque control of induction motors based on discrete space vector modulation using adaptive sliding mode control was proposed by Ben Salem and Derbel [3], the results shows the effectiveness and the robustness of the DTC- discrete SVM adaptive sliding mode control of induction motors. The variations of induction motor parameters is shown by Ben Salem and Derbel in their subsequent two publications namely Performance Analysis of DTC-SVM Sliding Mode Controllers-Based on Estimator of Electric Motor Speed Drive [4], and DTC-SVM Based Sliding Mode Controllers with Load Torque Estimators for Induction Motor Drives [5] respectively. The advantages of this direct torque control method is improved efficiency and fast response of torque in dynamic conditions [6, 7].

2. Types of induction motor

A typical three phase Induction motor consists of two parts namely stator and rotor, the outer part is called stator having coils supplied with three phase AC current to produce a rotating magnetic field. The inside rotating part is called rotor attached to the output shaft that is gives the useful torque produced by the rotating magnetic field. The stator is made up of stack of steel laminations of 0.35-o.5 mm thick with slots similar to a stator of a synchronous machine. The Coils are placed in the slots to form a three or single phase winding. **Figure 1A** shows the stator stampings with slots of induction motor **Figure 1B** shows the stator of induction motor [6].

The rotors of induction motors are of two types namely squirrel cage rotor and slip ring rotor. The squirrel cage rotor is made up of punched laminations with (0.35 to 0.5) mm thick steel core with rotor slots. Aluminum bars are molded in the slots instead of winding. End rings short circuit the aluminum bars at each side [6]. **Figure 2** shows the squirrel cage rotor.

The slip ring rotor or wound rotor has windings like the stator and at the end of each phase the winding is connected to a slip ring. There are three slip rings and

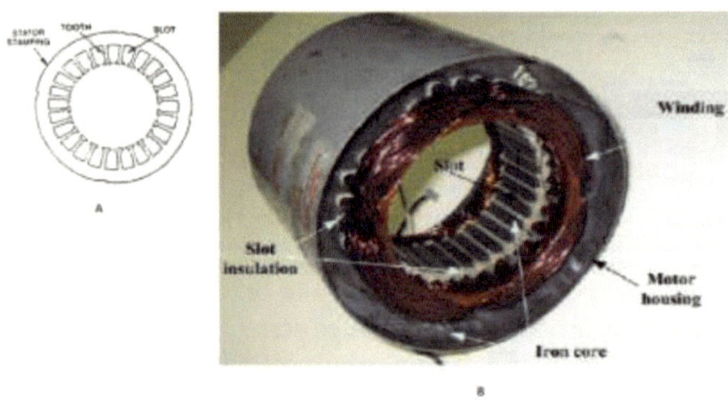

Figure 1.
(A) Stator stamping with slots of induction motor. (B) Stator of Induction Motor.

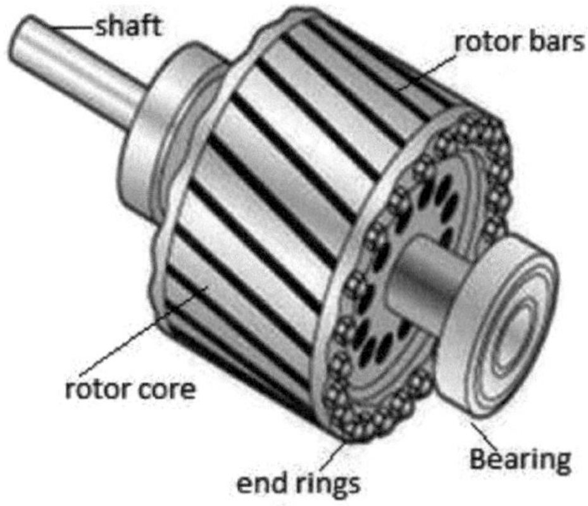

Figure 2.
Squirrel cage rotor.

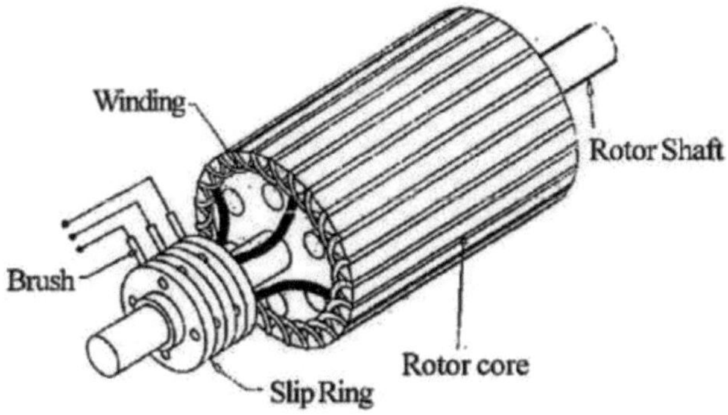

Figure 3.
Slip ring rotor.

three brushes through which three resistances can be connected in three phase star configuration for reducing starting current and speed control as well as increasing the torque [6]. **Figure 3** shows the slip ring rotor.

3. Classification of IM control strategies

The various IM control techniques are classified in to scalar and vector control methods. The general classification of IM control strategies [8, 9] which are based on the variable frequency control is shown in **Figure 4**.

3.1 Scalar control

The various scalar control methods are as follows 1. Stator voltage control 2. frequency control 3. Voltz/Hertz (V/F) control 4. Rotor Voltage control 5. Changing the number of poles. Out of these scalar methods, V/F control method is the best scalar control method. It can able to adjust the speed of the Induction motor by controlling the amplitude and frequency of the stator voltage of induction motor, the ratio of stator voltage to frequency should be kept constant so that, it is called as V/F control of induction motor drive. The vector control is preferred over scalar control methods due to the following disadvantages of scalar methods 1. Control of

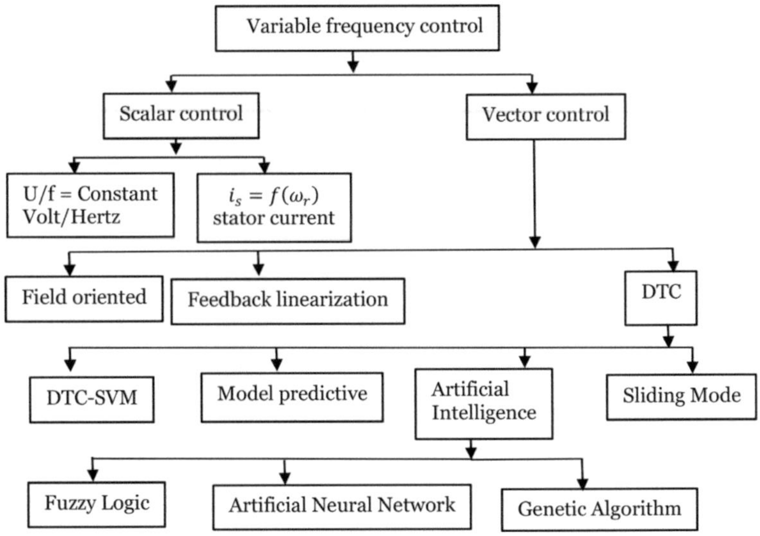

Figure 4.
General classification of control strategies of induction motor.

Voltage/Current/frequency magnitude is based on steady state equivalent circuit model which ignores transient conditions. 2. Coupling of torque and flux exists, and they are functions of frequency and voltage which leads to sluggish dynamic responses [6].

3.2 Vector control methods

3.2.1 Field oriented control (FOC)

The FOC method is implemented based on the analogy of controlling a DC motor. It does not guarantee an exact decoupling of the torque and flux in dynamic and steady state operations. The full information about motor state variable and load torque is required for controlling the IM. The relationship between regulated value and control variables is linear only for constant rotor flux amplitude. The current controllers, coordinate transformations and a PWM algorithm are required. For direct FOC, flux estimator is required. In indirect FOC mechanical speed sensor is needed. This method is very sensitive to rotor time constant.

3.2.2 Direct torque control (DTC)

The DTC is one of the high performance control strategies for the control of AC machine. In a DTC drive applications, flux linkage and electromagnetic torque are controlled directly and independently by the selection of optimum inverter switching modes of operation. To acquire a faster torque output, low inverter switching frequency and low harmonic losses in the model, the selection is made to restrict the flux linkages and electromagnetic torque errors within the respective flux and torque hysteresis bands. The required optimal switching vectors can be selected by using the optimum switching voltage vector look-up table. This can be obtained by simple physical considerations involving the position of the stator-flux linkage space vector, the available switching vectors, and the required torque flux linkage.

The torque is controlled by the stator current component I_{sq} in the classical vector control strategy of FOC according to the Eq. (1)

$$M_e = p_b \frac{m_s}{2} \frac{L_M}{L_r} \Psi_r I_{sq} \tag{1}$$

The motor torque is expressed by rotor flux magnitude Ψ_r and stator current component I_{sq} as given in the Eq. (2) and this equation is written as:

$$M_e = p_b \frac{m_s}{2} \frac{L_M}{L_r} \Psi_r I_s sin\delta \tag{2}$$

The Eq. (2) is transformed into the Eq. (3)

$$M_e = p_b \frac{m_s}{2} \frac{L_M}{L_r L_s - L_m{}^2} \Psi_s \Psi_r \ sin\ \delta_\Psi \tag{3}$$

where δ is the angle between rotor flux vector and stator current vector and δ_Ψ is the angle between rotor and stator flux vectors. The torque value depends on the magnitude of stator and rotor flux as well as the angle δ_Ψ. For FOC methods, the angle δ is considered whereas angle δ_Ψ is considered for DTC techniques.

The vector diagram of IM is shown in **Figure 5**.

From the motor voltage Eq. (5) for the omitted voltage drop on the stator resistance, the stator flux can be expressed.

From the mathematical model of IM, the electromagnetic torque equation is given in the Eq. (4)

$$M_e = p_b \frac{m_s}{2}\ Im\left(\Psi_s^* I_s\right) \tag{4}$$

Taking into consideration the fact that in the cage motor the rotor voltage equals zero and the electromagnetic torque Eq. (4), the following Eq. (5) is derived.

$$U_{sK} = R_s I_{sK} + \frac{d\Psi_{sK}}{dt} + j\Omega_K \Psi_{sK} \tag{5}$$

$$\frac{d\Psi_s}{dt} = U_s \tag{6}$$

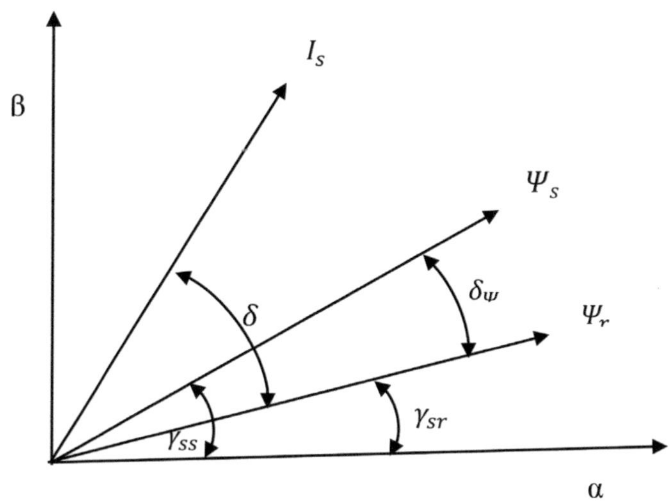

Figure 5.
Vector diagram of induction motor.

Taking into consideration the output voltage of the inverter in the above Eq. (6) it can be written as

$$\Psi_s = \int_0^t U_v dt \tag{7}$$

where

$$U_v = \begin{cases} \frac{2}{3} U_{dc} e^{\,j(v-1)\pi/3} & v = 1 \dots 6 \\ 0, \dots \dots \dots \dots \dots \dots & v = 0, 7 \end{cases} \tag{8}$$

The Eq. (7) describes eight voltage vectors which correspond to possible inverter states. These vectors are shown in **Figure 6**. There are six active vectors U1 to U6 and two zero vectors U0, U7.

It can be seen from the Eq. (7) that the stator flux directly depends on the inverter voltage Eq. (8). By using one of the active voltage vectors the stator flux vector moves to the direction and sense of the voltage vector. Stator flux changes direction for the cycle sequence of the active voltage vectors. Inherently the rotor flux of IM moves slowly but the stator flux could be changed immediately. In DTC methods the angle δ_Ψ between stator and rotor flux is varied to control the torque. By adjusting the stator voltage, stator flux could be controlled in simple way. The above consideration and equations could be used in the analysis of classical DTC techniques and SVM-DTC methods. In the classical DTC method the control plane is divided for the six sectors shown in **Figure 7**, that are defined as:

$$\gamma_{ss} \in \left(-\frac{\pi}{6}, +\frac{\pi}{6}\right) \tag{9}$$

$$\gamma_{ss} \in \left(+\frac{\pi}{6}, -\frac{\pi}{2}\right) \tag{10}$$

$$\gamma_{ss} \in \left(+\frac{\pi}{2}, +\frac{5\pi}{6}\right) \tag{11}$$

$$\gamma_{ss} \in \left(+\frac{5\pi}{6}, -\frac{5\pi}{6}\right) \tag{12}$$

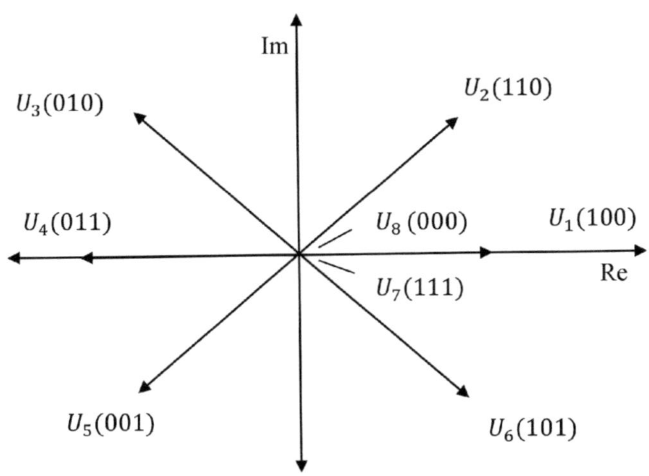

Figure 6.
Inverter output voltage represented as space vectors.

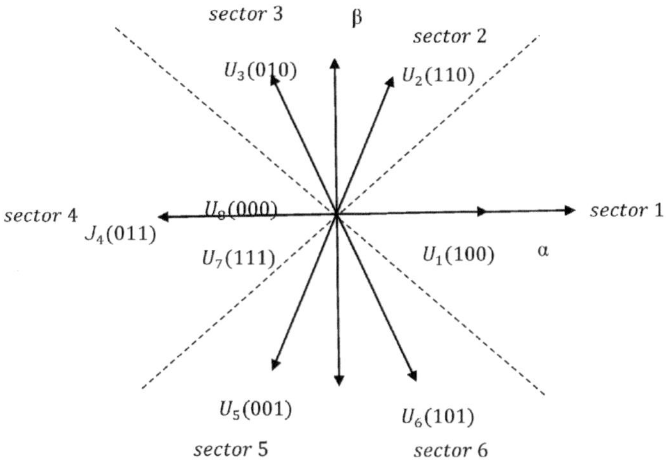

Figure 7.
Sectors in classical DTC.

$$\gamma_{ss} \in \left(-\frac{5\pi}{6}, -\frac{\pi}{2}\right) \tag{13}$$

$$\gamma_{ss} \in \left(-\frac{\pi}{2}, -\frac{\pi}{6}\right) \tag{14}$$

In order to increase magnitude of the stator vector in sector 1, the following voltage vectors $\mathbf{U_1}$ $\mathbf{U_2}$ $\mathbf{U_6}$ are selected. Conversely to decrease, $\mathbf{U_3}$ $\mathbf{U_4}$ $\mathbf{U_5}$ are selected. The stator flux is not changed when any one of the zero vectors $\mathbf{U_0}$ or $\mathbf{U_7}$ is applied. The solving of integration in Eq. (7) is stopped.

To increase the motor torque, the voltage vectors $\mathbf{U_2}$ $\mathbf{U_3}$ $\mathbf{U_4}$ are selected and for decreasing the torque $\mathbf{U_1}$ $\mathbf{U_5}$ $\mathbf{U_6}$ are selected. The switching **Table 1** is constructed based on the above considerations.

I. Takahashi and T. Nogouchi proposed the control scheme for Direct torque control and it is block diagram is shown in **Figure 8**.

The reference signals such as stator flux amplitude Ψ_{sc} and the electromagnetic torque M_c are compared with the estimated flux amplitude $\hat{\Psi}_s$ and electromagnetic torque $\hat{M}_e$ values respectively. The error values such as e_Ψ and torque e_M are sent to the hysteresis controllers. The appropriate voltage vector from the switching table is selected by the digitized output variables d_Ψ, d_M and the stator flux position sector $\gamma_{ss}(N)$. The power switches in the inverter are controlled by the pulses S_A, S_B, S_C which are generated form the switching table.

According to the Eq. (15) the torque and flux errors are calculated.

S_Ψ	S_m	S_1	S_2	S_3	S_4	S_5	S_6
1	1	U_2	U_3	U_4	U_5	U_6	U_1
	0	U_7	U_0	U_7	U_0	U_7	U_0
	−1	U_6	U_1	U_2	U_3	U_4	U_5
0	1	U_3	U_4	U_5	U_6	U_1	U_2
	0	U_0	U_7	U_0	U_7	U_0	U_7
	−1	U_5	U_6	U_1	U_2	U_3	U_4

Table 1.
Optimum switching table.

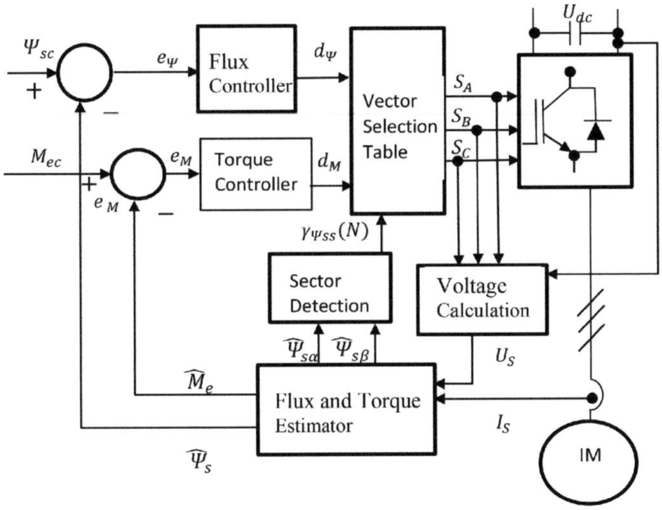

Figure 8.
Block diagram of DTC scheme.

$$\varepsilon_{\Psi_s} = \frac{\hat{\Psi}_s - \Psi_{sc}}{\Psi_{sN}} 100\% \tag{15}$$

3.2.3 Problems in conventional DTC

Despite its simplicity and robustness, the conventional DTC control has major drawback. The use of hysteresis controllers causes high ripples in the flux and electromagnetic torque at low speeds. It results in undesirable mechanical vibrations and acoustic noise, and subsequently leads to degradation of the machine performances. Thus the variable switching frequency and current distortions could detoriate the quality of the output power. The negligence in the calculation of stator resistance leads to problems at low speed. Moreover, the practical implementation of nonlinear components of the hysteresis type needs low sampling period. Many DTC strategies are developed based on the principle of instantaneous torque and stator flux regulation in order to rectify the drawbacks of classical DTC. The direct determination of the inverter control signals from the switching table is implemented [9].

4. Types of control strategies for torque ripple reductions in DTC

The DTC control strategies are divided into two groups: 1.Typical 2.Modern control strategies. They are classified into few other control techniques such as space vector modulation (SVM-DTC), modified switching table (m-DTC), Artificial Neural Network controller based (ANN-DTC), Fuzzy Logic controller based (FLC-DTC), Genetic algorithm based (GA-DTC), Model predictive controller based (MPC-DTC), Sliding mode based (SMO-DTC) [9, 10] as shown in **Figure 9**.

4.1 Modified DTC

It covers modification in switching table and/or injection of dithering signals. Few attempts are made to avoid the drawbacks in convention DTC either by implementing dither signal injection method or modified switching table method.

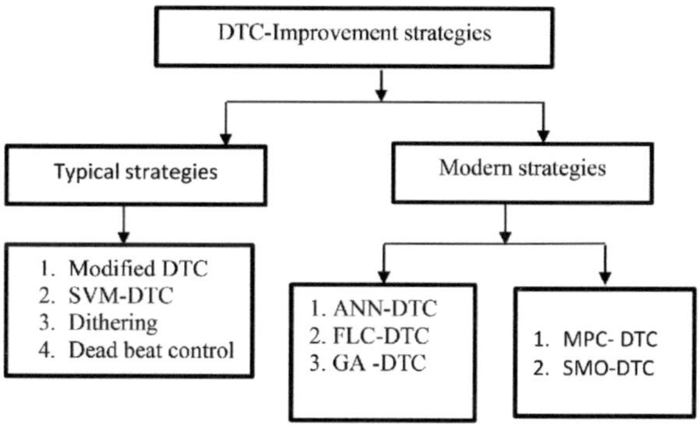

Figure 9.
Classification of DTC improvement strategies.

4.1.1 Modification in switching tables

The modifications are carried out in the DTC- basic switching table with the objective of improving starting and overload conditions which enable all the voltage vectors are applied in appropriate sequence. They are implemented by two methods namely 1. Six sector table and 2. Twelve sector table respectively. The zero voltage vectors are selected from the switching **Table 1** during starting and very low speed conditions and results in flux level reduction due to the drop in stator resistance [11].

4.1.1.1 Modified switching table

4.1.1.1.1 Improvement in switching table

In conventional DTC, the states v1 and v4 vectors are not used. Depending on if the position is in its first 39 degrees or in its second ones, they could increase or decrease the torque. It leads to modify the switching table and use the modified DTC. In the modified DTC, the vectors v3 and v6 are not used. The reason is the ambiguity in flux instead of torque as if it was in conventional DTC [12].

4.1.1.1.2 Modified classical DTC

By applying zero voltage vectors V0, V7 for the states of decreasing in torque, **Table 1** is modified accordingly. The inertia of the motor is reduced when zero voltage vectors are applied, torque ripple is reduced. It is more suitable than the percent given by applying the voltage vectors in **Table 1** for the torque decrease states. **Table 2** illustrates this modification [12].

Voltage vectors	Classical DTC	DTC with changes of zones
V_1	+30° to −30°	0° to −60°
V_2	+90° to +30°	+60° to 0°
V_3	+150° to +90°	+120° to +60°
V_4	−150° to +150°	+180° to +120°
V_5	−90° to −150°	−120° to −180°
V_6	−30°to −90°	−60° to −120°

Table 2.
Modified switching table with 6-sectors.

In both classical DTC and modified DTC there are two states per sector that present a torque ambiguity, so they are never used either. Instead of six sectors, the stator flux locus is divided into twelve sectors. Then all six active states will be implemented per sector. Consequently, the idea of the twelve sector modified DTC [13] is introduced. The tangential voltage vector component is very small and consequently its torque variation will be small as well. Based on this fact, the technique of small torque increase instead of torque increase is implemented [10, 11].

4.1.2 Dither signal injection

Feedback signals should not be delayed in order to maintain maximum possible switching frequency. Due the presence of isolation amplifier, Hall effect transducer and other components, the delay is made inevitably. By introducing the dither signal at very high frequency, the effect due to delay could be compensated. Normally these dither signals are triangular waves at double or triple the sampling frequency of the system. This dithering technique minimizes the torque ripple to 30% compared to conventional DTC method [14].

The frequency of the dither signal is selected well above the cutoff frequency of the system so that its presence could not be detected in the output. When the system parameters are not exactly known and not alterable, the method of instantaneous injection of dither signal is robust to noise in measurements. The inherent delay in signal transduction, data acquisition system and computation leads to low switching frequency which would result in increased torque and flux ripples. The dithering signal injection is implemented to improve the switching frequency of inverter. The appropriate magnitude and frequency of dither signals which are injected in torque and flux errors could minimize torque ripples and acoustic noise level in the drive [15].

4.1.3 Deadbeat control

In the inverter operation to avoid a short circuit in the DC-link, only one switch is turned on at a time. During the transistor switching signals, a delay time must be inserted and as a result the transistors stops to conduct. The dead-time TD is presented for the transistors T_1, T_2 for the two control signals SA+, SA- respectively. Most of the transistors take 1-3 μs duration of dead-time. The safe operation of the inverter is guaranteed by this delay time but it results into a serious distortion in the output voltage. Consequently there is a loss of control momentarily, where a deviation in output voltage from the reference voltage is observed. It is repeated for every switching cycle, so it has significant impact on the control of the inverter and this is known as dead-time effect. The inverter has nonlinear characteristics due to the dead-time and voltage drop on the switching devices. So the compensation algorithms are required in the control strategies [8] as shown in **Figure 10**.

The dead-beat DTC scheme is based on the technique, forcing the magnitude of torque and stator flux to attain their reference values in one sampling period. It is achieved by synthesizing a suitable stator voltage vector applied from Space Vector Modulation (SVM). In this approach the changes in the magnitude of torque and flux over one sampling period are calculated from the motor equations. To get the command value of stator voltage vector in stationary coordinates, a quadratic equation is solved [11].

The flux estimation is crucial part in the sensor less control strategies. The algorithm used for this is sensitive on the calculation accuracy of the inverter output voltage. From the switching signals, the voltages are reconstructed. **Figure 11** shows

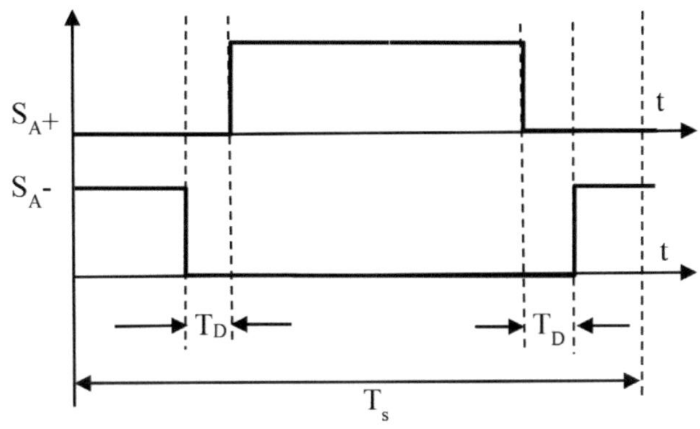

Figure 10.
Dead-time effect in PWM inverter.

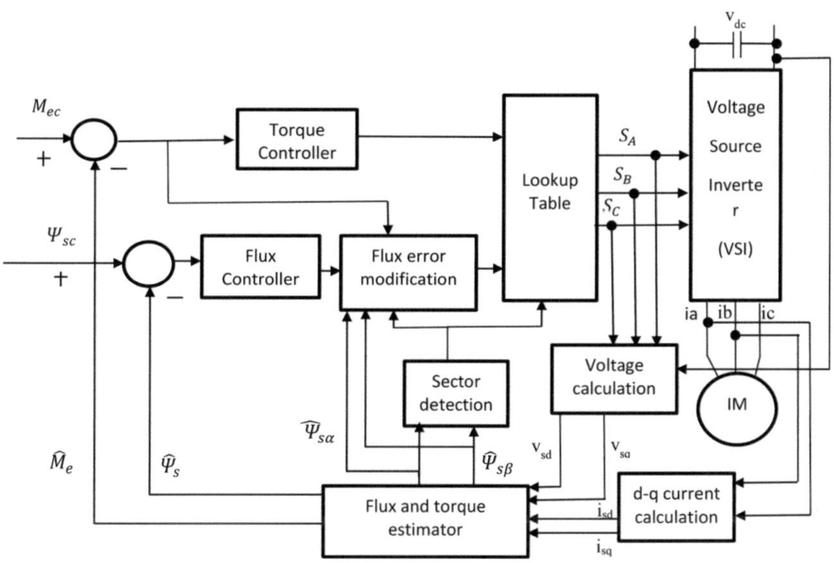

Figure 11.
DTC control with modification of flux error status block.

the DTC control with modification of flux error status block [13] dead-time compensation algorithm is significant in this SVM-DTC method [8].

The torque and flux ripple are reduced when the switching frequency of the inverter is maintained constant and greater than the sampling frequency [11].

4.1.3.1 Constant switching frequency approach

For general purpose IM drives, PI-DTC is an appropriate solution in a very wide power range. It is suited to very fast torque and flux controlled drives because of its short sampling time which is required by the switching table DTC schemes [10].

The stator resistance influences the estimation accuracy of stator flux. The characteristics of both torque and flux control loops are affected by error in estimation of stator flux. A new strategy in MATLAB/SIMULINK model is implemented with modified flux error block which resulted in getting quick response [13].

The largest tangential vector to the circular flux locus is produced by an optimized voltage vector. This voltage vector is switched and held to achieve a fast rate of change of angle Δδsr. The optimized voltage before being it is fed to the lookup table; its selection is done by modifying the flux error status [13].

4.1.4 SVM-DTC

The main difference between classical DTC and DTC-SVM (**Figure 12**) control methods lies in which the control algorithm is being used for the calculations. Based on the instantaneous values, the classical DTC algorithm directly calculates the digital control signals for the inverter. In the DTC-SVM methods control algorithm calculations are based on averaged values whereas the switching signals for the inverter are calculated by space vector modulator. Based on voltage model, the flux estimator with reference flux is selected for the implementation DTC-SVM control structure in sensor less mode of operation [8].

4.1.4.1 SVM

The classical DTC has several disadvantages, among which the variable switching frequency and the high level of ripples are the prominent issues [16]. Further they lead to high-current harmonics and an acoustical noise and they detoriate the control performance at low speeds. The ripples are produced proportionally to the width of the hysteresis band. Due to the discrete nature of the hysteresis controllers, even for the reduced bandwidth values, the ripples are still present [16].

The inverter switching frequency is increased due to even very small values of bandwidths. The modifications in classical DTC strategy is done by including a vector modulator block, which produces space vector PWM technique (SVM) and it is used to implement the voltage vector with a fixed frequency of inverter switching. The switching table and hysteresis controllers are replaced with PI controllers to control the stator flux and the torque [13].

The disadvantages of DTC-SVM with conventional PI controllers are as follows 1. Sensitivity to variation in system parameters and 2. Inadequate rejection of external disturbances and 3. Load changing conditions. These disadvantages are overcome by replacing the conventional PI controllers by intelligent controllers such as adaptive fuzzy-PI or FLC. These intelligent controllers ol more robust against the external disturbances and parameter variations [13].

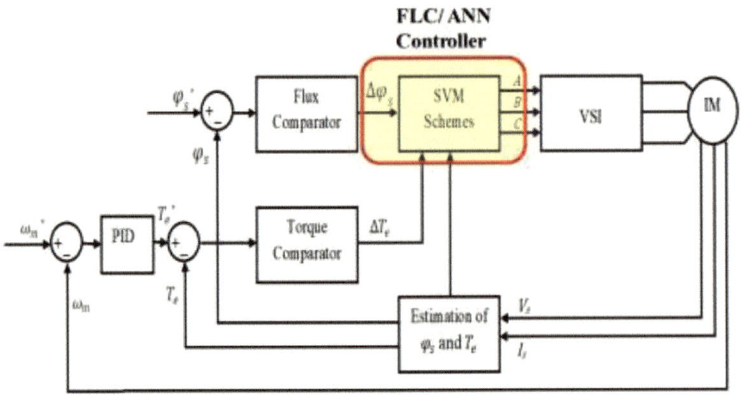

Figure 12.
Block diagram of FLC/ANN controller for DTC-SVM scheme for induction motor.

4.1.4.2 Types of SVM-DTC

The DTC-SVM methods have several following classes namely:

1. PI controllers based DTC-SVM.

2. Predictive/dead-beat based DTC-SVM.

3. Fuzzy logic and/or neural networks based DTC-SVM.

4. Variable-structure control (VSC) [8] based DTC-SVM.

The use of PI controller for torque control of induction motor drives is to overcome an overshoot during startup and to minimize steady state error. The PI controllers provide feedback signals to the system.

In voltage model based stator flux estimation, the pure integrator is replaced with LPF to eliminate the problem of saturation and integration drift due to the DC offsets which are present in the sensed currents or voltages. The LPF introduces the phase and magnitude errors of stator flux estimation which affects the selection of voltages vector and electromagnetic torque response, thereby it deteriorate the performance of DTC drive. To overcome the LPF problems, closed loop of stator flux estimation is implemented [17].

In MRAS, to estimate the rotor speed, PI controller is used and this controller takes more time to tune the proportional and integral gain to obtain the estimated target speed. The MRAS is based on rotor speed, rotor flux and stator current thereby it eliminates the need of PI controller [18].

The effective integration of SVM technique with any n-level multilevel inverter fed DTC drive is achieved by using a fractal based space-vector DTC algorithm. The current THD performance is improved at higher level of DTC drives under transient, steady state and speed reversal operating conditions. Without any significant modification, this strategy could be adapted at any n-level inverter fed DTC controller [19].

4.1.4.3 Proposed artificial intelligent schemes for DTC-SVM

The Space Vector PWM (SVPWM) is a technique used for solving the switching losses in the power converter. The SVPWM is operated in a symmetrical way, so the switching state of each sector is predefined. In this proposed scheme, the initial values to the DTC controller have been fixed based on the induction motor rating. Then the estimation of DTC parameters is found and it is fed to the reference to the Hybrid Asymmetric Space Vector PWM (HASVPWM) controller.

Traditional PWM techniques consist of two signals called carrier signal and reference signal for generating the PWM pulses. If any distortions in the reference signal (i.e error signal) may produce miscellaneous pulses, which will affect the performance of the converter. But SVPWM technique is purely based on estimating the voltage magnitude and its angle for pulse generation. In this, three phase voltages V_{abc} are converted into V_d,V_q and V_0 using abc-dq0 theory. This method will make the estimation of the sector angle and voltage magnitude easier. In traditional SVPWM, each sector denotes 60^0 angles and totally it has six sector and two reference vector in its implementation. Even though the estimation is done for every sector accuracy of generating the pulses is lagging due to the higher range of sector angle and minimum switching sectors. For an example, if estimated sector angle is 55^0, then the switching pattern in sector 1 is selected for the PWM generation. But V_2 vector is also having different switching pattern and that may also well suited for the same estimated sector angle 55^0. In order avoid such difficult situations, a HASVPWM is used for controlling the DTC drive which reduces torque ripples, switching losses and improved power quality.

The implementation of HASVPWM is similar to the SVPWM technique. In general, three phase Voltage source inverters (VSI) have eight distinct switching losses, where state 1 to 6 are active states, 0 and 7 are inactive switching states. In HASVPWM, asymmetric voltage vectors are represented as V_{ni}, V_{nj} and V_{nk} where n = 1,2,3,4,5,6 and four quadrants. HASVPWM has two non-zero vectors (V_1 and V_2) and two zero vectors (V_0 and V_{24}) in each vector will be used for the vector 15^0. Hence this HASVPWM have 24 sectors and it is shown in **Figure 13**.

Major portion in HASVPWM is to removal of mismatching pulses which will be done by comparing the HASVPWM pulses with the traditional SVPWM pulse. The mismatching pulses are removed by calculating its rise time (T_r) and fall time (T_f) of the mismatching pulses with magnitude. Then the same magnitude of pulse with same instant is added. For mismatching pulse removal. This logic avoids the mismatching pulses in the output and reduce the switching losses in the VSI based DTC drive. In this proposed system, intelligent control methods such as Fuzzy Logic Control (FLC) and Artificial Neural Network (ANN) are utilized to find the suitable sector for continuous operation. They are also efficient than the classical control techniques which are utilized to find suitable sector for the continuous operation.

4.1.4.3.1 Fuzzy logic controller for HASVPWM

The proposed hybridization process is performed by the combination of continuous ASVPWM and fuzzy operated Discontinuous ASVPWM technique. Finally, the mismatching pulses of both PWM techniques are applied to control the inverter. Pulse mismatching technique helps to reduce the active region of the switch and achieve the optimal input pulse to the inverter. Pulse mismatching technique helps to reduce the active region of the switch and achieve the optimal input pulse to the inverter. The optimal hybrid pulse reduces transition time of inverter switch and improves operating performance of the inverter. The Fuzzy rules help to select the optimal switching sector for discontinuous modulation. If there is more number of sectors in the hexagon, it allows more degrees of freedom which help to find the optimal reference voltage and angle. The Fuzzy logic system describes to what degree the rule applies, while the conclusion assigns a fuzzy function to each of one or more output variables. These Fuzzy Expert Systems allow more than one conclusion per rule.

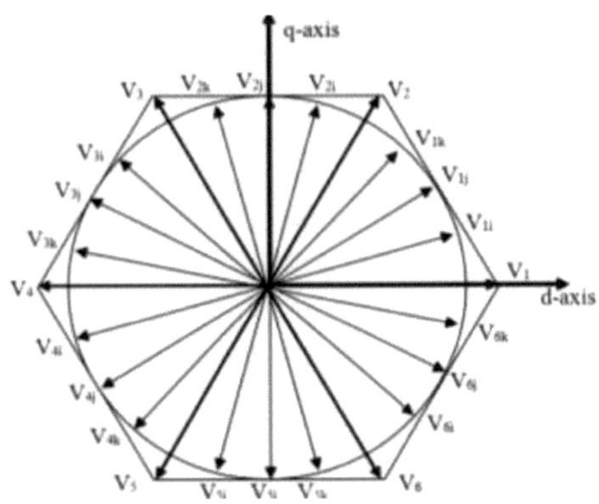

Figure 13.
Structure of HASVPWM hexagon.

CE\E	NB	NS	ZE	PS	PB
NB	NB	NB	NS	NS	ZE
NS	NB	NS	NS	ZE	PS
ZE	NS	NS	ZE	PS	PS
PS	NS	ZE	PS	PS	PB
PB	ZE	PS	PS	PB	PB

Table 3.
Fuzzy Logic Rules to select suitable sector in HASVPWM.

The linguistic labels are divided into five groups. They are: NB-Negative Big; NS- Negative Small; ZE-zero; PS-Positive Small; PB-Positive Big. Each of the inputs and the output contain membership functions with all these five linguistics.

The set of rules in a fuzzy expert system is given in **Table 3** and corresponding input and membership function values are indicated in **Figure 8**.

The simulation model of DTC with HASVPWM scheme is developed using MATLAB software Simulink tool. The fuzzified inputs and defuzzified outputs are shown in **Figures 14–16** respectively. Consider a case, when the sector angle estimated from the SVPWM calculation as shown in **Figure 14** is equal to -165^0, it means negative big (NB) as mentioned in **Table 3**. And change in the sector angle at the next instant is about -110^0, it represent the negative small (NS). Then the corresponding fuzzy output is NB, which is mentioned in **Figure 16**.

4.1.4.3.2 Artificial neural network controller for HASVPWM

The DTC control can also be achieved with HASVPWM using Artificial Neural Network (ANN) control. ANN is nonlinear model that is easy to use and understand compared to statistical methods like Fuzzy logic. Compare with Fuzzy Logic, ANN has an ability to learn from the previous trained data. Hence, the major advantage of

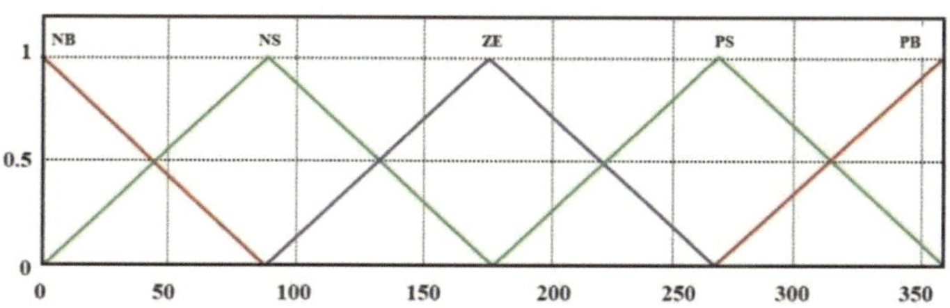

Figure 14.
Degree input to FLC.

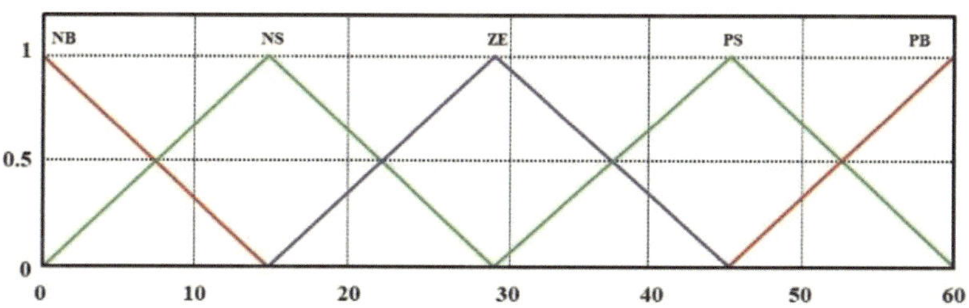

Figure 15.
Change in degree error input to FLC.

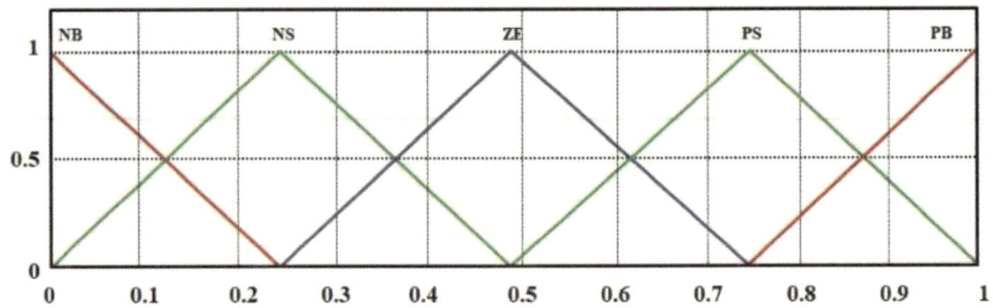

Figure 16.
Fuzzy output Modulation index for HASPWM.

ANN is to train a system with large amount of data sets. The output performance will depend upon the trained parameters and the data set relevant to the training data. In this proposed scheme, ANN is used to estimate the suitable sector of HASVPWM.

ANN is used to determine the sector number for the estimated value of θe. There are total of 24 sectors, each sector of 15 degree. Again three layers of neurons are used but with a 5–4-1 feed forward configuration as shown in **Figure 17**. The Input layer is of log sigmoid transfer function, hidden layer is of hyperbolic tangent sigmoid function and the output layer is of linear transfer function. Levenberg - Marquardt back propagation based training method is used for train the neurons. As soon as the training procedure is over, the neural network gives almost the same output pattern for the same or nearby values of input. This tendency of the neural networks which approximates the output for new input i.e. angle theta since sector selection is purely based on theta.

The following **Figure 17** shows the structure of Neural Network (NN) which is utilized in the proposed ANN controller for DTC-SVM scheme for induction motor.

The Step by step procedure for NN Algorithm is given below:

Step 1: Initialize the input weight for each neuron.

Step 2: Apply the training dataset to the network. Here X is the input to the Network and Y_1, Y_2 and Y_3 are the output of the network.

Step 3: Adjust the weights of all neurons.

Step 4: Determine Sector Angle for SVPWM.

Compare with Fuzzy logic control, ANN control in HASVPWM is able to identify the suitable voltage vector and its angle for minimizing the torque ripple and PEC losses and THD, maximizing DTC capabilities under various operating conditions like speed reversal, loading conditions etc. The effectiveness of ANN-HASVPWM in DTC scheme is predicted by comparing ANN with the Fuzzy based

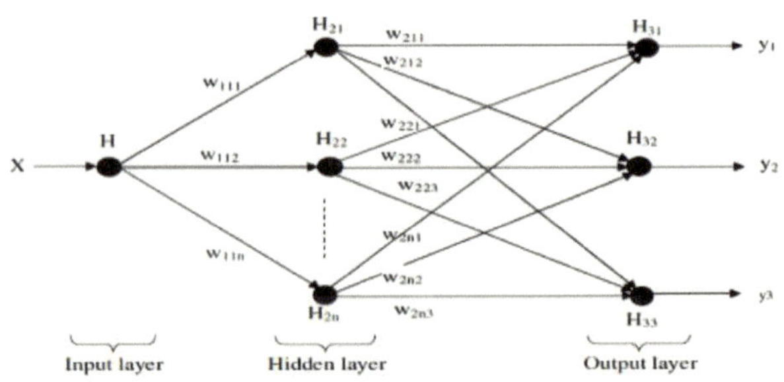

Figure 17.
The structure of network utilized in the proposed technique.

Control strategies	Torque ripple factor (%)
Proposed FLC controller for HASVPWM FOR DTC-SVM scheme	5.1
Proposed ANN controller for HASVPWM FOR DTC-SVM scheme	4.5

Table 4.
Comparison of control strategies in induction motor.

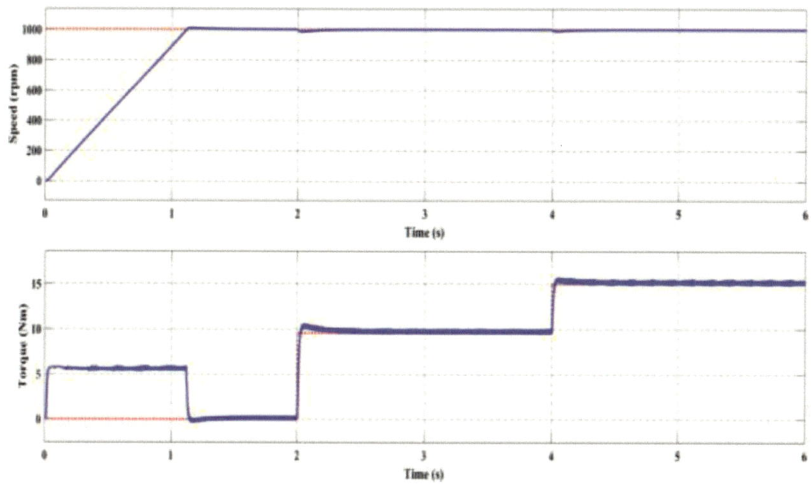

Figure 18.
Speed and torque output for FLC based DTC-SVM of IM with HASVPWM.

HASVPWM scheme. The results for ANN based HASVPWM scheme to DTC controller under the same loading conditions, it shows that torques ripple, switching loss and harmonic content reduction is expected. The comparative simulation results are clearly presented and shown in **Table 4**.

4.1.4.3.3 Simulation results and discussion

The two proposed schemes namely 1.Fuzzy Logic Controller (FLC) for DTC-SVM and 2.Artificial Neural Network (ANN) controller for DTC-SVM respectively for IM. Has been simulated using MATLAB version R2009a and the results are compared and shown in **Table 4**. Both of the proposed scheme methods uses HASVPWM. The parameters of IM used in the simulation are given in the appendix.

The torque ripple can be calculated by using the relation.

$$\text{Torque Ripple Factor} = (\text{Peak to Peak torque})/\text{Rated torque} \tag{16}$$

The simulation results of FLC for DTC-SVM of IM with HASVPWM is shown in **Figures 18** and **19**.The simulation results of ANN for DTC-SVM of IM with HASVPWM is shown in **Figures 20** and **21**.

From **Figure 19** (Torque ripple waveform) it is inferred that the torque ripples oscillates from 9.5 Nm (Minimum) to 10.1 Nm (maximum) for the given Reference torque of 10 Nm.

Torque Ripple factor (%) as per Eq. 23 is given by = ((10.01–9.5)/10) $\times$ 100 = 0.51/10$\times$100 = **5.1.**

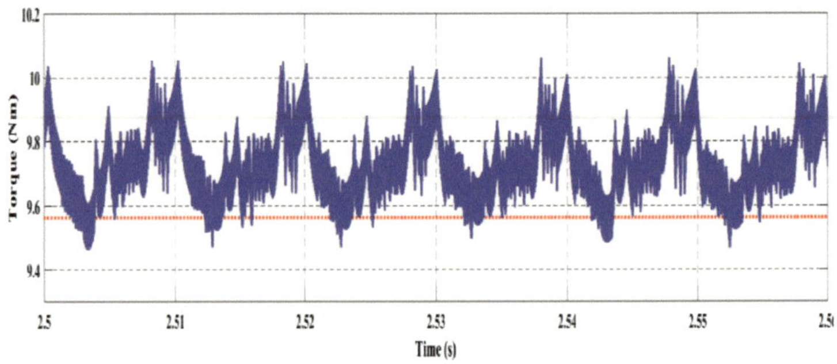

Figure 19.
Torque ripples for FLC based DTC-SVM of IM with HASVPWM.

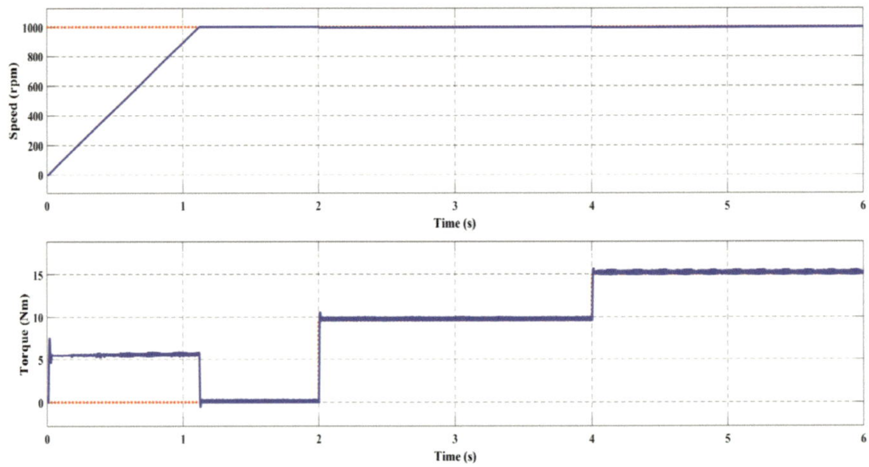

Figure 20.
Speed and torque output for ANN based DTC-SVM of IM with HASVPWM.

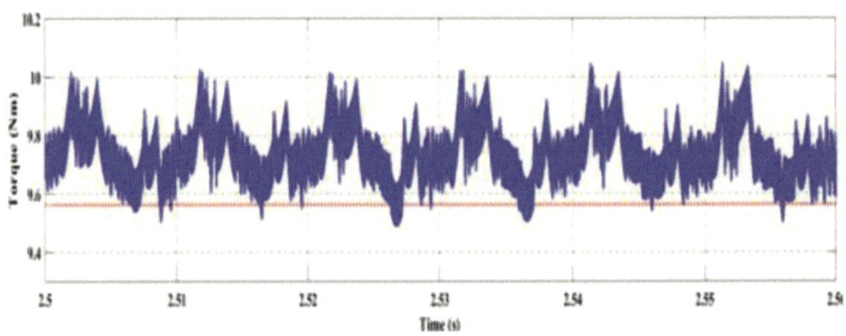

Figure 21.
Torque ripples for ANN based DTC-SVM of IM with HASVPWM.

From **Figure 21** (Torque ripple waveform) it is inferred that the torque ripples oscillates from 9.55 Nm (Minimum) to 10 Nm (maximum) for the given Reference torque of 10 Nm.

Torque Ripple factor (%) as per Eq. 23 is given by = ((10–9.55)/10) × 100 = 0.45/10×100 = **4.5.**

Figure 22 shows the comparison of Torque Ripple of FLC and ANN based DTC-SVM of IM with HASVPWM at 1000 rpm.

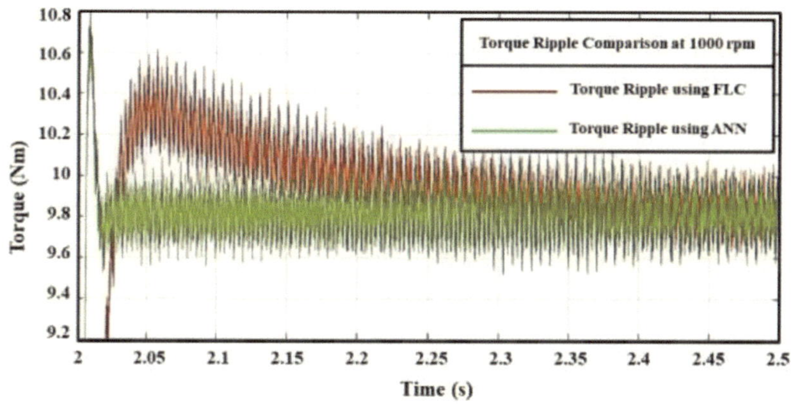

Figure 22.
Torque ripple comparison of FLC and ANN based DTC- SVM of IM with HASVPWM.

It is clear that variation in Torque ripples shown in **Table 4** is less in case of ANN and they can achieve a minimum torque ripple than other control techniques. It has been viewed that the discussed control strategy has helped in reducing the torque ripples. Thus, by using FLC and ANN based controller for DTC of IM, the ripples are reduced completely.

5. Modern strategies of DTC

5.1 Fuzzy logic control (FLC)

The limits of the torque hysteresis band are controlled by FLC. It entails a minimization of the torque ripples as well as an improvement of the dynamic performances of IM. The FLC selects the optimum bandwidth of the torque hysteresis in real time [9].

The fuzzified parameters such as torque error, stator flux errors, and stator flux angle are the input to FLC. The switching state of the inverter is a crisp value obtained as an output from the FLC [11].

A detailed classification and comparison of DTC strategies like SMC, FLC and ANN in terms of performance parameters of induction machine were discussed in [20].

The PI controller and FLC algorithm have been implemented for the three-phase induction motor and it is found that the proposed FLC scheme is better than the conventional DTC control with PI control in [21].

The effective DTC improvements are achieved by Fuzzy logic controller thereby the ripples in torque and flux are reduced, consequently secondary problems for the motor such as heating, mechanical vibration, aging are also rectified. The merits of the conventional are also preserved [22].

5.2 Artificial neural network (ANN)

A multilayer ANN allows to replace both hysteresis comparators and the selection table in classical DTC.

The ANN offers the following merit over classical DTC.

i. The complexity of the controller is reduced;

ii. The effects of motor parameter variations are minimized.

iii. The controller time response is improved.

iv. The robustness of drive is improved [11].

For electric vehicle applications, the FLC and ANN based monitoring systems for the DTC controlled induction motor drive was implemented to detect a very small change in performance parameters [23].

Acknowledgements

We are thankful to the management of following institutes, the Department of Electrical and Electronics Engineering, PSG Institute of Technology and Applied Research, Coimbatore, Tamilnadu, India, and the Department of Mechatronics Engineering, Nehru Institute of Engineering and Technology, Coimbatore, India, for their encouragement, support and facilities provided for our book chapter work.

Nomenclature

f_c	crossover frequency
i_d	d-axis current
L_{dm}	d-axis magnetizing inductance
L_d	d-axis self-inductance
V_d	d-axis voltage
ρ	derivative operator
T_e	develop electromagnetic torque
d	direct or polar axis
DTC	direct torque control
ω_r	electrical speed
i_f	equivalent permanent magnet field current
L_s	equivalent self-inductance per phase
λ_d	flux linkage due d axis
λ_q	flux linkage due q axis
λ_{dm}	flux linkage due to rotor magnets linking the stator
B	friction
FLC	fuzzy logic controller
J	inertia
k_i	integral control gain
T_L	load torque
ω_{rated}	motor rated speed
T_m	motor torque
P	number of poles
I_m	peak value of supply current
λ_f	PM flux linkage or field flux linkage
k_p	proportional control gain
i_q	q-axis current
L_{qm}	q-axis magnetizing inductance
L_q	q-axis self-inductance
V_q	q-axis voltage
q	quadrature or interpolar axis

T_{ref}	reference motor torque
θ_r	rotor position
ω_m	rotor speed
L	self-inductance
L_{ls}	stator leakage inductance
R_s	stator resistance
i_a, i_b, i_c	three phase currents
V_a, V_b, V_c	three phase voltage

A. Appendices

A.1 Appendix 1: parameters of 3 phase squirrel cage induction motor

RATED POWER	5.4 HP
RATED VOLTAGE	400 V
RATED SPEED	1430 RPM
RATED TORQUE	26.7 Nm
RATED CURRENT	8.5 A
STATOR RESISTANCE	1.405 Ω
ROTOR RESISTANCE	1.395 Ω
STATOR INDUCTANCE Ls	0.005839H
ROTOR INDUCTANCE Lr	0.005839H
MAGNETIC FLUX	0.1827 weber
NO. OF POLES	4
MOMENT OF INERTIA	0.0131 kgm^2
FRICTION FACTOR	0.002985 Nms/rad

Author details

Adhavan Balashanmugham[1]*, Maheswaran Mockaisamy[2]
and Sathiyanathan Murugesan[1]

1 Department of Electrical and Electronics Engineering, PSG Institute of Technology and Applied Research, Coimbatore, Tamil Nadu, India

2 Department of Mechatronics Engineering, Nehru Institute of Engineering and Technology, Coimbatore, India

*Address all correspondence to: adhav14@gmail.com

IntechOpen

© 2020 The Author(s). Licensee IntechOpen. This chapter is distributed under the terms of the Creative Commons Attribution License (http://creativecommons.org/licenses/by/3.0), which permits unrestricted use, distribution, and reproduction in any medium, provided the original work is properly cited.

References

[1] Takahashi, Noguchi T. A new quick response and high-efficiency control strategy of an induction motor. IEEE Transactions on Industry Applications., Vol.22, No.5, pp.820–827, 1986. DOI: 10.1109/TIA.1986.4504799

[2] Depenbrock M. Direct self control (DSC) of inverter fed induction machine. IEEE Transactions on Power Electronics. 1988; 3(4):420–429. DOI: 10.1109/63.17963

[3] Ben Salem, F. and Derbel N., 'Direct torque control of induction motors based on discrete space vector modulation using adaptive sliding mode control', International Journal of Electric Power Components and Systems, 2014, 42, (14), pp. 1598–1610.

[4] F. Ben Salem and N. Derbel, 'Performance Analysis of DTC-SVM Sliding Mode Controllers-Based Parameters Estimator of Electric Motor Speed Drive', Mathematical Problems in Engineering, Volume 2014, Article ID 127128, http://dx.doi.org/10.1155/2014/127128, 2014.

[5] Ben Salem,F. and Derbel,N., 'DTC-SVM Based Sliding Mode Controllers with Load Torque Estimators for Induction Motor Drives', Chapter 14 of the Book: Applications of Sliding Mode Control, Studies in Systems, Decision and Control 79, Springer Science+ Business Media Singapore 2017, pp. 269–297

[6] P.C. Krause,O. Wasynczuk, S. D. Sudhoff "Analysis of Electric Machinery and Drive Systems", IEEE Press, A John Wiley & Sons, Inc. Publication Second Edition, 2002.

[7] H. C. Stanley, "An Analysis of the Induction Motor", AIEE Transactions, Vol. 57 (Supplement), pp. 751–755, 1938.

[8] Marcin Żelechowski. Space Vector Modulated – Direct Torque Controlled (DTC–SVM) Inverter–Fed Induction Motor Drive, Dissertation, Warsaw University of Technology; Warsaw, 2005. p. 1–175.

[9] Najib El Ouanjli et al. Modern improvement techniques of direct torque control for induction motor drives - a review. Protection and Control of Modern Power Systems. Springer Open; 2019. p. 1–12. https://doi.org/10.1186/s41601-019-0125-5.

[10] Vojkan Kostić et al. Expermental Verification of Direct Control Methods for Electric Drive Application. Facta Universitatis. Series: Automatic Control and Robotics; Vol. 8, No.1, 2009. pp. 111–126.

[11] M.Vasudevan. Improved Direct Torque Control Strategy with Ripple Minimization for Induction motor drive. Anna University: Chennai; 2006. p. 39–87.

[12] Eng.Ahmed Hassan Adel et al. Torque Ripple Reduction in Direct Torque Control of Induction Motor Drives by Improvement of the Switching Table. Journal of Multidisciplinary Engineering Science and Technology; Vol. 1 Issue 5, 2014. p. 1–6.

[13] Madhuri D. Kulkarni, Vivek D. Bavdhane. Quick Dynamic Torque Control in DTC-Hysteresis-Based Induction Motor by Using New Optimized Switching Strategy. International Journal of Innovations in Engineering Research and Technology; Vol. 2, Issue 7, 2015. p. 1–11.

[14] Jaya N. Tattea, Mohoda, S. B. Performance Improvement of Induction Motor by Using Direct Torque Control Technique. International Journal of Development Research; Vol. 07, Issue 10, 2017. pp.15901–15905.

[15] RK.Behra, SP.Das Improved direct torque control of induction motor with dither injection, Sadhana; Vol. 33, Part 5, 2008. p. 551–564.

[16] Cherifi Djamila, Miloud Yahia. Direct Torque Control Strategies of Induction Machine: Comparative Studies. Direct Torque Control Strategies of Electrical Machines. Intechopen p. 1–23. DOI: http://dx.doi.org/10.5772/intechopen.90199.

[17] Rajendran R. Certain Investigations and Realization of SVM-DTC of Induction Motor Drives Using FPGA. Anna University, Chennai; 2012. p. 46–66.

[18] Hassan Farhan Rashag et al. DTC-SVM Based on PI Torque and PI Flux Controllers to Achieve High Performance of Induction Motor. Research Journal of Applied Sciences, Engineering and Technology 7(4); 2014. p. 875–891.

[19] Vinod B. R., Shiny G. A Multilevel Inverter fed Direct Torque Control Strategy for an Induction Motor using PI Controllers. International Journal of Engineering and Advanced Technology; Volume-7 Issue-4, 2018. p. 90–99.

[20] Najib El Ouanjli et al. Modern Improvement Techniques of Direct Torque Control for Induction Motor Drives - A review, Protection and Control of Modern Power Systems. 2019, 4:11,https://doi.org/10.1186/s41601-019-0125-5, pp. 1–12.

[21] J. Jeyashanthi, M. Santhi. Performance of Direct Torque Controlled Induction Motor Drive by Fuzzy Logic Controller. CEAI; Vol.22, No.1, 2020, pp. 63–71.

[22] Najib El Ouanjli et al. Improved DTC strategy of doubly fed induction motor using fuzzy logic controller. Elsevier Energy Reports 5, 2019, pp. 271–279.

[23] Abarna J, Velnath R. Modeling of Three Phase Induction Motor with DTC Drive Fault Analysis using Fuzzy Logic. International Journal of Recent Technology and Engineering (IJRTE); ISSN: 2277–3878, Vol. 9, Issue-1, May 2020. pp. 1196–1202.